Frontiers in Pain Science

(Volume 1)

(New Non-opioid Analgesics: Understanding Molecular Mechanisms on the Basis of Patch-clamp and Quantum-chemical Studies)

Authored by:

Boris V. Krylov, Ilia V. Rogachevskii, Tatiana N. Shelykh & Vera B. Plakhova

I.P. Pavlov Institute of Physiology Russian Academy of Sciences, St. Petersburg, Russia

Frontiers in Pain Science

Volume # 1

New Non-opioid Analgesics: Understanding Molecular Mechanisms on the Basis of Patch-clamp and Quantum-chemical Studies

Author: Boris V. Krylov, Ilia V. Rogachevskii, Tatiana N. Shelykh & Vera B. Plakhova

eISSN (Online): 2542-9175

ISSN (Print): 2542-9167

eISBN (Online): 978-1-60805-930-0

ISBN (Print): 978-1-60805-931-7

General:

1. Any dispute or claim arising out of or in connection with this License Agreement or the Work (including non-contractual disputes or claims) will be governed by and construed in accordance with the laws of the U.A.E. as applied in the Emirate of Dubai. Each party agrees that the courts of the Emirate of Dubai shall have exclusive jurisdiction to settle any dispute or claim arising out of or in connection with this License Agreement or the Work (including non-contractual disputes or claims).

2. Your rights under this License Agreement will automatically terminate without notice and without the need for a court order if at any point you breach any terms of this License Agreement. In no event will any delay or failure by Bentham Science Publishers in enforcing your compliance with this License Agreement constitute a waiver of any of its rights.

3. You acknowledge that you have read this License Agreement, and agree to be bound by its terms and conditions. To the extent that any other terms and conditions presented on any website of Bentham Science Publishers conflict with, or are inconsistent with, the terms and conditions set out in this License Agreement, you acknowledge that the terms and conditions set out in this License Agreement shall prevail.

Bentham Science Publishers Ltd.
Executive Suite Y - 2
PO Box 7917, Saif Zone
Sharjah, U.A.E.
Email: subscriptions@benthamscience.org

**BENTHAM
SCIENCE**

CONTENTS

PREFACE

In 1897, at a meeting of the Society of Russian Physicians, Ivan Pavlov predicted that the last stage of the life sciences would be the physiology of the living molecule. Nowadays the last stages of molecular approaches are theoretical quantum-chemical calculational techniques and experimental patch-clamp method which really can describe the behavior of living molecules. An attempt of combined application of quantum-chemical calculations and the patch-clamp method to investigation of the nociceptive system is presented in this volume. The crosstalk between drug substances and membrane receptors is conducted in the language of molecules. The behavior of single molecules upon their ligand-receptor binding should be investigated at physiologically adequate conditions during development of new analgesics. The requirement of physiological adequacy was always taken into account when the authors tried to explain the background mechanisms governing the effects of powerful analgesics. This approach makes it possible to elucidate how the chemical structure of labile attacking molecules should be finely tuned to provide effective binding to their membrane receptor. The authors hope that this review will open a new perspective to application of molecular methods in the drug design of pain relievers. The urgent need for the development of novel analgesics is dictated by the lack of safe and effective drugs in this field of medicine, especially when the pain becomes intolerable and incurable. The arsenal of practical medicine includes an array of analgesics, which have to be applied basing on the severity of pathological conditions of the organism. Step 1 of the World Health Organization analgesic ladder consists of non-opioids, administered with or without adjuvants depending on the type of pain. Step 2 comprises step 1 agents plus opioids which can relieve mild to moderate pain. Step 3 involves step 2 agents with addition of opioids for moderate to severe pain relief. It is a matter of common knowledge that administration of opioid substances results in irreversible adverse side effects in humans. The major objective of the authors is to solve this underlying problem by creating novel analgesics which could replace opioids in clinical practice, while remaining completely safe.

This book presents our main result in elucidation of the physiological role of a novel membrane signaling pathway involving the opioid-like receptor coupled to slow sodium channels ($Na_v1.8$) via Na^+,K^+-ATPase as the signal transducer. This pathway is distinct from and additional to the known mechanism of the opioidergic system functioning that involves G proteins. Activation of the opioid-like receptor further triggering the signaling pathway directed towards $Na_v1.8$ channels provides the effectiveness and safety of our novel analgesic which is potent enough to relieve severe pain otherwise relieved exclusively by Step 3 opioids.

It is nowadays almost inevitable for reviewers of scientific material in the field of nociception to make excuses for omissions. We are sincerely sorry for not having been able to discuss all the findings in physiology of nociception and in practical medicine that would have merited attention. To include all would have defeated the purpose of this volume by making it grow out of all proportions.

This book presents an informative and valuable for physiologists and clinicians overview of primary molecular mechanisms involved in functioning of the peripheral nociceptive system. This material can be used in courses given to students specializing in physiology, psychology, and medicine, as well as to physicians training in neurology, neurosurgery, and psychiatry. The principles presented in the current volume may also be of interest to molecular biologists engaged in the drug design.

Boris V. Krylov
Ilia V. Rogachevskii
Tatiana N. Shelykh
Vera B. Plakhova
I.P. Pavlov Institute of Physiology
Russian Academy of Sciences
St. Petersburg,
Russia

Frontiers in Pain Science

Frontiers in Pain Science, 2017, *Vol. 1*, 3-61

CHAPTER 1

Introduction and Methodology

Abstract: Discovery of $Na_V1.8$ channels has opened a new perspective to study the mechanisms of nociception. A remarkable feature of these channels is their ability to be modulated by binding of various endogenous and exogenous agents to membrane receptors coupled to $Na_V1.8$ channels. The behavior of their activation gating system was patch-clamp recorded and analyzed by the Almers' limiting slope method. It was established that opioid-like membrane receptors could control the functioning of $Na_V1.8$ channels. A novel role in this mechanism is played by Na^+,K^+-ATPase, which serves as the signal transducer instead of G proteins. Switching on the opioid-like receptors one can selectively decrease the effective charge of $Na_V1.8$ channel activation gating device. As a result, only the high-frequency component of nociceptive membrane impulse firing is inhibited. This is the component that transfers nociceptive information to CNS.

The three units involved in the described membrane signaling cascade (opioid-like receptor $\rightarrow$ Na^+,K^+-ATPase $\rightarrow$ $Na_V1.8$ channel) are potential targets for novel analgesics. Investigation of this mechanism of nociceptive signal modulation is of major importance not only for fundamental physiology but also for clinical medicine.

Keywords: Impulse firing, Limiting slope procedure, $Na_V1.8$ channels, Na^+,K^+-ATPase, Nociception, Opioid-like receptor, Patch-clamp method.

PHYSIOLOGY OF PRIMARY SENSORY CODING

The universal language of the brain is the language of nerve impulses. In the 1920s Edgar Adrian was the first who discovered that discharge frequency of an afferent fiber innervating feline mechanoreceptors increased as a consequence of an increase in the stimulus intensity. The input-output function of the primary afferent fiber describes the relationship between the stimulus intensity and the number and frequency of evoked action potentials [1]. Different forms of energy are transformed by the nervous system into different sensations of sensory modalities. Five major sensory modalities have been recognized since ancient times: vision, hearing, touch, taste, and smell. In 1844 Johannes Müller advanced his "laws of specific sense energies" [2]. He proposed that modality was a property of the sensory nerve fiber. Each fiber is activated by a certain type of stimulus because different stimuli activate different nerve fibers. In turn, the nerve

Boris V. Krylov, Ilia V. Rogachevskii, Tatiana N. Shelykh, Vera B. Plakhova

fibers make specific connections within the nervous system, and it is these specific connections that are responsible for specific sensations. A unique stimulus that activates a specific receptor and therefore a particular nerve fiber was called an adequate stimulus by Charles Sherrington [3]. In 1967 Vernon Mountcastle advanced the idea of the brain as a "linear operator" [4, 5]. It means that the input-output functions of sense organs should be congruent with psychophysical functions relating the magnitude of the stimulus to the sensation. For instance, the function of central pathways mediating simple sensory events in the somatosensory system is thought to conserve the presentation of a stimulus dictated by the peripheral sensory apparatus. Said differently, one could assign for each discriminable quality of sensation a specific set of nerve fibers whose excitation would express that one quality (modality) and no other. The alternative view stated that quality was a matter of the pattern or of the spatio-temporal distribution of excitation in a whole array of fibers. As a result, the "labeled line" theory was opposed to the alternative "pattern" theory (see review) [6].

The "sixth" sensory modality, *i.e.* pain, up to now attracts special attention of physiologists and clinicians. It is difficult to overestimate the significance of attempts to control the mechanisms of pain sensation in order to achieve practical results regarding chronic pain relief in humans. The first steps in this direction have been done by researchers who laid the foundations of nociception as one of the most important branches of sensory physiology.

Alfred Goldscheider (1920) [7] was the first to advance the idea that the pain was not modality-specific but rather evoked by an additional excitation of any sense organ.

Ivan Pavlov (1927) [8] showed how the brain could be trained, through repetition, to invoke certain reactions in certain circumstances. Pavlov distinguished between food stimulations which called out the reaction of salivation and electric current noxious stimulations which called out the defense reaction. Destructive (noxious) stimuli provoke the defense reflex. Food calls for a positive reaction – grasping of the substance and eating it. Pavlov has shown that the defense reflex of skin is second in importance to the food reflex. An animal exposed simultaneously to an electric current acting upon his skin and to a food stimulus would respond not with defense but with food reaction. These findings show that mediation of nociceptive signals does not strictly obey the "labeled line" law. This "line" is under control of some other physiological processes of living organism.

Investigating the physiological nature of sleep Pavlov stated that sleep, or inhibition, prevented undue fatigue of the cortical elements, allowing them, after they had been subjected to noxious stimulation, to recover their normal state.

Inhibition is occurring all the time, even in a seemingly alert animal, but it exists only in scattered areas of the cortex. When it irradiates from these areas over the entire brain, the animal falls asleep. In Pavlov's words, "internal inhibition in the alert state of the animal represents a regional distribution of sleep which is kept within bounds by antagonistic nervous process of excitation" (Pavlov, Conditioned reflexes, p. 253) [8]. Pavlov has demonstrated that the nature of the stimulus itself is less important than the inhibition associated with it. "As there is practically no stimulus of whatever strength that cannot, under certain conditions, become subjected to internal inhibition, so also there is none which cannot produce sleep" (Pavlov, Conditioned reflexes, p.252) [8]. He mentions an instance in which a powerful electric shock applied to the skin was used as a conditioned alimentary stimulus that totally relieved pain (see also [9]).

There are three main consequences of Pavlov's findings. The first one is that his results corroborate the "pattern" theory, because as it was mentioned above, the "noxious labeled line" could be easily disrupted by signals of other modalities in an alert organism. The second consequence is the suggestion that nociceptive signals can be controlled somewhere at spinal and/or supraspinal levels. And finally, nowadays we can predict that endogenous substances which should control pain sensation on the molecular level are expressed in human brain.

A starting point of any sensation is the reception of signals evoked by activation of specialized sensory receptors (including nociceptors) providing information to CNS. Nociceptors inform us mainly about harmful external and internal stimuli or about tissue injury. Pain is the perception of an aversive or unpleasant sensation that originates from a damaged region of the body. Our "sixth sense" is a vitally important sensory experience that warns us on danger. Modern findings concerning the relationship between perception of pain and mechanisms of functioning of nociceptors show that any nociceptive perception involves an interconnection and elaboration of sensory inputs and pathways. Highly subjective and complicated nature of pain makes it difficult to diagnose and treat a number of chronic pain phenomena.

A noxious stimulus activates the nociceptor fiber by the fundamental mechanism of excitation of living cell. It is well known that nerve excitation evoked by mechanical stimulation results in production of gradual receptor current in primary receptors [10, 11] or generator current in secondary receptors [12] that elicits the single action potential or trains of nerve impulses. Insights into neural mechanisms for fine coding of tactile information in humans come from the works of Ake Vallbo and his colleagues who have systematically studied mechanoreceptors innervating the human hand skin. On the basis of information obtained on alert subjects they have proven that even single extra action potential is of major importance, as it informs us about changes in the intensity of tactile stimulus [13, 14]. The authors managed to record the impulse activity of single peripheral fibers that innervate cutaneous mechanoreceptors. Besides, the subjects reported a change in their sensations in response to an increase of the adequate stimulus force [15]. According to the results of questioning the subjects, the authors succeeded in dividing the psychophysical scale of sensations into five levels. In spite of a certain scatter, these levels correlated to the number of simultaneously recorded action potentials in the afferent fiber. If touching the receptor field was so weak that nerve impulses either did not arise at all or only one action potential appeared, the subjects reported the absence of sensations. As the stimulus force was increased, the sensation became vague. One action potential corresponded to this threshold, but it appeared with a greater probability. A very weak sensation correlated with appearance of one or two action potentials, a weak one was noted when three or four appeared, and when the sensation was moderate, five or six, more rarely, nine action potentials were observed in the afferent fiber. These results were supported by other authors [16]. It was shown that the maximal number of nerve impulses arising in the fibers innervating slowly adapting skin receptors in response to adequate stimulation was five. In this case, subjective judgment of the sensation level was linearly dependent upon the number of action potentials. The results of these investigations seem to have allowed the final solution of the problem concerning physiological significance of each nerve impulse in the train arising in the nerve fiber.

The next discovery of great importance was made due to studies carried out on both humans and experimental animals which demonstrated that an increase in nerve firing frequency immediately arising from "damaging" mechanical, thermal, or chemical stimuli resulted in pain sensation. A direct correlation has been found when human psychophysical responses to noxious heat stimuli were compared with the receptive properties of nociceptive afferents recorded from anesthetized monkey [17 - 19] or alert humans [20 - 22]. These studies have also indicated that increased pain perception following tissue injuries can be explained by the corresponding change of the stimulus-response functions of nociceptive afferents [21 - 25]. Under normal physiological conditions, nociceptive signals are produced by intense stimulation of primary afferent sensory Aδ and C nerve fiber terminals by chemicals, heat, and pressure [26, 27]. In addition, Koltzenburg and Handwerker [28] found that response of a nociceptor increased with the velocity of a projectile stimulus. Similarly, C fiber nociceptors [17] and innoxious thermal receptors [29, 30] exhibit both a rate- and temperature-sensitive response to thermal stimuli. Though some kinds of stimuli applied to the skin are psychophysically painful, comparable mechanical forces applied gradually do not inevitably induce pain. Nonpainful stimuli could elicit instantaneous discharge with interspike intervals of less than 100 ms, *i.e.*, with instantaneous frequencies of more than 10 Hz [28], which indicates that a brief high-frequency burst of unmyelinated nociceptors should not be necessarily sufficient to induce pain. On the other hand, the mean number of action potentials evoked by a painful impact stimulus was around eight impulses. Some discrepancy between the nociceptor discharge and pain perception has also been observed for other mechanical [31, 32], thermal [20, 31], and chemical [33] stimuli. These studies made it possible to estimate that the average nociceptor discharge rates exceeding 0.5-1.0 Hz during a maintained stimulus were required to evoke painful sensations. This could mean that the temporal summation of a certain number of impulses in nociceptive afferents should be necessary for conscious perception of pain in humans, in spite of limited correlation between the frequency modulation of the discharge and perceived sensation [28]. The cited authors also found that monotonic increase of total nociceptor discharge following impact stimulation of increasing stimulus intensity was accompanied by corresponding increase of vasodilatation. In this case, the total number of action potentials is also the determinant for the magnitude of neurogenic vasodilatation after a single noxious stimulus. Such parallel changes of pain sensation and vasodilatation have been observed after thermal [34], electrical [35, 36], and chemical stimulation [37]. These results support the Pavlov's prediction that the nociceptive system is tightly coupled to visceral systems. An important conclusion concerning physiological complexity of pain sensations has been obtained in humans using intraneural microstimulation. It has been determined that the magnitude of pain evoked by electrical excitation of nociceptive afferents depends on the pattern characteristics of stimulation [38]. A pattern that mimicked the natural discharge of nociceptors and consisted of a dynamic high-frequency discharge that settled on a lower frequency was generally perceived as more painful than the same number of impulses delivered over the same time with a regular interstimulus interval. One of explanations of this phenomenon is that the decrease in firing frequency is due to desensitization of excitable membrane responsible for analog-impulse transformation in the nociceptor. On the other hand, the summation of pain sensations observed in experiments on humans has many features in common with slow increase of excitability of higher-order neurons studied in animals. It has been found that repetitive electrical stimulation of C fibers results in the progressive increase of excitability of spinal cord neurons [39 - 42], although there are different mechanisms that could account for this observation. The most important "station" that nonlinearly processes nociceptive signals is the Melzack's "gate".

In the early 1960s neurophysiological studies provided evidence that stimulation of low-threshold myelinated primary afferent fibers decreased the response of dorsal horn neurons to unmyelinated nociceptors, whereas blockade of conduction in myelinated fibers enhanced the response of dorsal horn neurons. Firing of certain spinal cord neurons may therefore not simply be regarded by the level of activity in nociceptive afferent input, but also by the balance of activity between unmyelinated nociceptors and myelinated afferents not directly related with pain. This idea was introduced by Patrick Wall and Ronald Melzack as the "gate control theory" [43]. Modulation of pain may be realized due to interactions of four classes of neurons in the dorsal horn of the spinal cord: (1) unmyelinated nociceptive fibers (C fibers), (2) myelinated nonnociceptive afferents (Aα, Aβ), (3) projection neurons, whose activity results in pain sensation, and (4) inhibitory interneurons. Inhibitory interneurons are spontaneously active and normally inhibit projection neurons, thus reducing the pain intensity.

The gate control theory introduced the idea that pain perception was sensitive to the levels of activity in both nociceptive and nonnociceptive afferent fibers. The theory elegantly explains the fact that nociceptive signals can also be modulated at successive synaptic relays along the central pathway. It is excited by myelinated nonnociceptive afferents but inhibited by unmyelinated nociceptors. Nociceptors thus have both direct and indirect effects on projection neurons.

It can be concluded that relatively unspecialized nerve cell endings that initiate the pain sensation have their cell bodies in the dorsal root ganglia (or in the trigeminal ganglion) which send their dendrite processes to the periphery and their axons to the spinal cord or brainstem. Faster-conducting Aδ nociceptors respond to dangerously intense mechanical and/or mechanothermal stimuli. Other unmyelinated polymodal nociceptors respond to thermal, mechanical, and chemical stimuli. There are three major classes of nociceptors in the skin: Aδ mechanosensitive nociceptors, Aδ mechanothermal nociceptors, and polymodal nociceptors, the latter being specifically associated with C fibers. Nociceptive dendrites begin to discharge only when the strength of a stimulus reaches high enough levels.

As it was shown above, sensory signals normally originate at a dendrite of pseudobipolar sensory neuron in the segmental dorsal root ganglia (DRG). Its membrane also might generate firing activity, but this is sparse in intact animals and its sensory consequences are likely to be minor in normal cells [44, 45]. After nerve injury, however, a discharge originating ectopically within DRG is greatly augmented and can be a major contributor to neuropathic diseases and chronic pain [44, 46 - 56]. This discharge is critically dependent on subthreshold membrane potential oscillations; oscillatory signals that reach the threshold trigger low-frequency trains of intermittent spikes. Ectopic firing may also enter a high-frequency bursting mode, particularly in the event of neuropathy [57].

In addition to increasing the firing frequency, stronger stimuli also activate a greater number of receptors, so that the intensity of a stimulus is also encoded in the size of the responding receptor population. This way of coding by the number of simultaneously activated receptors is of great importance when receptor population activity is studied. Nevertheless, herein we shall solely consider coding processes in a single receptor. Electrically excitable nociceptive membrane will be further shown to be able to perform the function of an analog-impulse convertor which encodes an analog impulse signal, the receptor (or generator, or synaptic) current, into the train of action potentials.

Summarizing the above, one can conclude that primary sensory coding of noxious stimuli is based on specific physiological responses of the human organism. Immediate reaction of the peripheral nervous system is manifested in the increase in impulse firing of nociceptors, although this rule may have some exceptions. The pattern theory of the mechanisms of nociception can be analyzed quantitatively using the Hodgkin-Huxley formalism describing the behavior of membrane ion channels. Our working hypothesis is based on the postulate that the nociceptive neuron membrane should incorporate ion channels which critically control the excitability of this neuron and, therefore, inhibitory modulation of their gating machinery should result in pain relief. Molecular mechanisms by which diverse chemical, mechanical, and thermal noxious stimuli depolarize free sensory endings and trigger an increase in action potential firing could generate a pathological state of the human organism. Correction of this pathology (pain relief) is usually achieved by reducing the number of active ion channels controlling nerve excitation. A different approach is applied herein. Our aim is to perform a physiologically adequate modulation of the gating device (a part of molecular structure of the channel) of slow $Na_v1.8$ sodium channel which is responsible for coding of nociceptive signals (see below).

ROLE OF ION CHANNELS IN PRIMARY SENSORY CODING

There are three main manifestations of primary sensory impulse firing in the nociceptive system: frequency code, numerical code, and spike frequency adaptation. Activating particular ion channels and analyzing their gating machinery characteristics, one could understand molecular mechanisms of processing of nociceptive signals. In other words, each ion channel plays its own specific physiological role in the process of analog-impulse transformation in nociceptors. As it was discussed above, noxious signals manifest themselves primarily in a significant increase of nociceptive membrane firing frequency. This is why to relieve pain one should control the process of analog-impulse transformation in the nociceptive membrane in such a way that would decrease the frequency of nerve impulses. This decrease should be the main effect of an analgesic drug. Antinociceptive action could also be achieved as a result of a decrease in the number of nerve impulses in a train or due to activation of the process of spike frequency adaptation in the nociceptive membrane. It will be demonstrated further that manifestations of nociceptive signal control are based on the ionic mechanism of modulation of slow sodium $Na_v1.8$ channels responsible for coding of nociceptive information. These channels represent the superfamily of sodium channels ($Na_v1.1 - Na_v1.9$). It is worthwhile to review the main physiological findings explaining the role of sodium channels in impulse coding. Behavior of the nociceptive system is determined by general laws of the information processing in the brain. To understand them on the molecular level it is necessary to get an insight on structure, functions, and mechanisms of intermolecular coupling of sodium channels which, together with calcium channels, are the key molecules ensuring generation of the action potential. Finding a way to decrease the excitability of the nociceptive membrane is of major clinical importance because millions of people are suffering from chronic pain without many perspectives of its safe and effective treatment.

The sequence of events between an increase of magnitude of receptor (generator) current and changes in frequency of excitable membrane impulse firing may be complicated; it results in structural changes in voltage sensors of ion channels. These parts of the channel have a large charge or dipole moment, thus being able to change their position in space in response to transmembrane voltage step. Any ion channel has more than one conformational steady state. Each of these stable conformations represents a different functional state of the gating device. Transition of a channel between the closed and open states is referred to as gating. The gating machinery of sodium channels consists of activation and inactivation gates. Initially, two approaches to describe the process of voltage gating have been suggested: one of them pictures permanent dipoles that flip-flop between two possible positions, while the other one considers charged particles belonging to the gating device which are trapped in the membrane and redistribute themselves between its two positions on external and internal sides. In both of these models the transition should obey first-order kinetics. The two-state, constant dipole model provides a useful starting point for quantitative analysis of the gating machinery. The dipole moment of a voltage sensor changes in a more complicated multi-state manner in a macromolecule, the overall orientation of which in the membrane does not change [58]. The voltage sensor belongs to voltage-gated ion channel molecule. Alan Hodgkin and Andrew Huxley were the first to describe the ionic mechanisms of generation of action potentials of the squid *Loligo forbesi* axon membrane [59 - 62]. Slightly later, Alan Hodgkin and Richard Keynes [63] put forward the concept of the "membrane ion channel". Aminoacid sequence of such a protein molecule was obtained for the first time only thirty years later [64].

The simplest process of generation of nerve impulses was predicted to be based on activation of two populations of ion channels, only one of them being voltage-dependent [65]. This minimal set of ion channels has been found in the membrane of intermediate Ranvier node that retranslates the impulse code produced by the first Ranvier node of an afferent fiber in warm-blooded animals. It involves the population of classical sodium channels ($Na_V1.1$) and the population of leakage (voltage-independent) channels [66, 67]. Classical $Na_V1.1$ channels and leakage current channels form a basic system which provides firing of the main "carrier" impulse frequency of the excitable membrane generator. Introduction of the third type of ion channels, classical potassium channels, results in an important change in the input-output function of the coding membrane. This function becomes negatively accelerated, which is a characteristic feature of a majority of receptors [6, 68, 69]. The other finest mechanisms regulating and tuning the coding membrane generator characteristics are based on activation of an additional population of voltage-dependent ion channels. Activation of slow potassium channels in the nerve fiber results in spike frequency adaptation [65]. This manifestation of impulse and numerical coding can also be achieved due to second-order properties of inactivation gating system of classical sodium channels [69]. It can be concluded that gating devices of ion channels participate in specific tuning of coding machinery. Thus, intra- or intermolecular modulation of gating devices by endogenous or exogenous agents is very important for functioning of the nociceptive system, which is determined by the mechanisms of impulse firing (see below).

Analysis of the nucleotide sequence of the sodium channel gene, as well as of the aminoacid sequence that it encodes, has revealed fundamental structural features of the channel. Each sodium channel consists of an α subunit of approximately 2,000 amino acids and a smaller auxiliary β subunit that modifies the properties of the α subunit. The α subunits, Na_V1, are encoded by a family of genes called SCN alpha. The human gene family includes SCN alpha genes numbered 1–11. The sodium channel α subunit proteins are numbered using a separate sequence which unfortunately does not entirely correspond to the numbering of the genes. The SCN10A gene product is called $Na_V1.8$ [70, 71]. It is this channel that plays a key role in encoding of nociceptive signals (see below). α subunits of Na_V1 channels consist of four roughly symmetrically arranged domains (DI-DIV) connected by intracellular loops. Each domain involves six α-helical segments (S1-S6) and segments S5 and S6 which participate in pore forming. This pore serves as the selectivity filter controlling passage of ions of a certain size across the membrane. The second fundamental insight into structural organization of the sodium channel is based on the observation that one of its putative membrane-spanning regions, called the S4 region, is highly conservative among sodium channels. It contains several positively charged aminoacid residues (lysines and arginines) that provide sensitivity to changes in voltage across the membrane. This region serves as the gating mechanism transforming voltage steps into the gating transition within the channel that opens the activation gate. The rest S1-S3 regions are probably involved in gating processes but there are no indications on uncompensated mobile charges involved in their structure [72 - 74]. Purification studies demonstrate that the sodium channel from mammalian brain is a complex of α (260 kDa), β1 (36 kDa), and β2 (33 kDa) subunits [75, 76]. β subunits are composed of an N-terminal extracellular immunoglobulin-like fold, a single transmembrane segment and a short intracellular segment. These subunits are thought to form heterodimeric and heterotrimeric complexes of a single α subunit and one or two β subunits in excitable cell membrane. Coexpression of β subunits modulates the kinetics and voltage dependence of sodium channel activation and inactivation, and extracellular immunoglobulin-like domains of β subunits serve as cell adhesion molecules that interact with matrix proteins and other cell adhesion molecules [76 - 78]. It is worthwhile to stress that the aminoacid structure of sodium channels is designed by the nature for effective interaction with the neighboring molecules. This coupling is of major importance for realization of physiological functions of sodium channels.

Selectivity Filter

To explain ionic selectivity, the original pore theory was put forward first by Loren Mullins [79] and later by George Eisenman [80] and Bertil Hille [81, 82] who proposed that channels should have a narrow region acting as a molecular sieve. At this selectivity filter, an ion sheds its surrounding water molecules and electrostatically interacts with charged aminoacid residues which line the walls of the channel. The most important structural feature of the selectivity filter is an oxygen-lined cavity roughly 3 by 5 Å in size. The observed selectivity (for Na^+ over K^+) was comparable both with a rigid selectivity filter, operating in a strict size-selection mode, and a flexible selectivity filter where the ion-coordinating residues could be pulled in by smaller ions and pushed out by larger ones [83]. Molecular dynamic simulations [84] showed that the selectivity filter was flexible/fluctuating. Ionic selectivity was shown to be determined by the balance of electrostatic forces and steric repulsion [85]. When ion-coordinating amino acids are packed as tightly as they are in the selectivity filter, the interactions between them must be taken into account [86]. In addition, it would be necessary to consider interactions of these amino acids with the environment outside the selectivity filter proper [87]. The selectivity filter can be modulated by just a few alkaloids, such as batrachotoxin, grayanotoxin, veratridine, and aconitine. Alkaloid-treated sodium channels display a more relaxed ionic selectivity than intact channels. While structurally distinct from one another, these highly hydrophobic compounds of low molecular weight do have certain common features, such as the bridged ring systems and esterified aromatic moieties found in batrachotoxin, veratridine, and aconitine [88]. In the case of batrachotoxin both the bridge and the esterified group are essential for its activity. An inner set of the residues Asp, Glu, Lys, and Ala (DEKA) formed the narrowest part of the selectivity filter [89]. Computer modeling and site-directed mutagenesis [90] made it possible to hypothesize that batrachotoxin could alter sodium channel ionic selectivity *via* interacting with its Phe1236 and Lys1237 residues. The pyrrole moiety of batrachotoxin forms aromatic electron cloud, which is why pyrrole becomes a good candidate for stacking with Phe1236. A hydrogen bond may then be formed between the amino group of the lysine side chain and the carbonyl group attached to the pyrrole group of batrachotoxin. Interactions between Lys1237 and batrachotoxin would distort the Lys1237 orientation, thus widening the selectivity filter. An increase of the selectivity filter diameter (from ~3.5 to ~3.9 Å) upon action of batrachotoxin has been suggested previously [91]. The discussed mechanism does not explain how batrachotoxin can recognize its targeting site in the selectivity filter. Hydrogen bonding and stacking interactions are too weak to account for a very high affinity ($K_d = 65 \div 85$ nM) of batrachotoxin binding. It can be thus predicted that electrostatic and ion-ionic interactions may play a key role in ligand-receptor binding (see below).

Alkaloids mentioned above were shown to bind to the narrowest internal part of the selectivity filter [89]. A remarkable distinctive feature of their effect is the fact that the sodium channel remains active, while the reversal potential of sodium current and ionic selectivity of the channel are crucially decreased. Several other substances can interact with a different part of the sodium channel (not with its selectivity filter) inhibiting its current without changing the reversal potential. Two of them are specific high-affinity sodium channel blockers (marine guanidinium toxins), saxitoxin (STX) and tetrodotoxin (TTX), which are widely used for investigation of voltage-gated sodium channels. TTX has been valuable in biochemical purification of the channels [92], in pharmacological characterization of different channel isoforms [93], in determination of the sodium channel density in cells [94], in discovery of slow sodium channels [95] and in elucidation of their role in nociceptive firing [96, 97].

These agents bind to toxin site 1, which is located on the outer vestibule of the channel [98, 99] and composed of Glu, Glu, Met, and Asp from domains I–IV, respectively. TTX and STX are suggested to bind to the channel in the same manner [89]. This idea is based upon the fact that the two toxins completely block the current, competitively inhibit each other in binding assays, are of similar size, and possess similar functional groups, including one or two guanidinium groups and a diol that are thought to be critically important for binding of either toxin. Furthermore, both toxins appear to block at the same depth within the electric field when a greater charge of STX is taken into account [100], and mutations in the outer vestibule seem to have qualitatively similar effects on affinities of both toxins [99, 101 - 103].

It is well known that the gating machinery of sodium channels can be modulated by drug binding. The most valuable results on drug modulation of the gating machinery were obtained in the last century by intensive investigation of the mechanisms of action of local anesthetics, which play a crucial role in inhibiting noxious signals. These mechanisms were demonstrated to be so sophisticated that existence of several receptor sites for local anesthetics with different properties on each sodium channel and of additional slow inactivated states of the channel had to be postulated by Boris Khodorov [104 - 106]. The hypothesis that the local anesthetics receptor of the sodium channel has at least three major states differing in their binding affinities made Bertil Hille to favor the "modulated receptor" hypothesis accounting for the action of local anesthetics [107]. This extremely fruitful idea is also relied on findings of Galina Mozhaeva and Aleksander Naumov [108 - 111]. In the 1970s – 1980s they were exploring the effects of ligand-receptor binding of some natural toxins on various characteristics of ion channels: conductance, selectivity, kinetics of activation and inactivation gating. These toxins were suggested to interact directly with aminoacid sequence of the channel in a dose-dependent manner which could be described by the Langmuir law. The modulated receptor hypothesis made it possible to explain a very high specificity of drug binding. However, when the Langmuir law cannot be applied to approximate the dose-effect function upon drug-channel interaction, it is necessary to apply the Hill equation for description of this nonlinear function.

The modulated receptor hypothesis can be treated more widely if we consider not only drug-channel interaction but also interactions of a drug with membrane receptors. Any membrane receptor is a protein structure which, upon binding of attacking molecules, switches on the sequence of cascade processes that further amplify and transduce the signal resulting from receptor activation to the genome and, in some cases, also to neighboring ionic channels (see below).

Inactivation Gating Device

Specific inhibition of inactivation process was firstly registered after internal perfusion of pronase in the squid axon. The agent preferentially removes inactivation while leaving the activation process intact, thus suggesting that some structures involved in inactivation process are accessible from the cytoplasmic side of the membrane [112]. Using the site-directed mutagenesis and patch-clamp method, sodium channel fragments with a "clipped" linker between DIII and DIV were later demonstrated to have impaired inactivation gating, implicating this loop in control on inactivation process [113]. Moreover, antibodies directed against residues 1491–1508 in the DIII–IV linker of neuronal sodium channels antagonized inactivation of single channels [114]. The implication of the DIII–IV linker has given rise to a working hypothesis that sodium channel inactivation proceeds through a "hinged-lid" mechanism, whereby linker residues serve as a molecular latch [115]. William Catterall has examined the details of three-dimensional hinged-lid mechanism by multidimensional NMR methods [75, 116]. The intracellular loop connecting domains III and IV of the sodium channel was shown to form a hinged lid, the critical residue Phe1489 being the main element occluding the intracellular mouth of the pore in the sodium channel during inactivation process [75]. Three amino acids Ile1488, Phe1489, and Met1490 form the IFM motif that belongs to a rigid α-helix. Thr1491, which is important for inactivation [117], and Ser1506, which is a site of phosphorylation and modulation by protein kinase C, are also involved in inactivation gating machinery. Further details of aminoacid architecture of this structure are discussed elsewhere [118 - 121].

Activation Gating Device

A commonly accepted hypothesis is that the main role in activation gating is played by S4 segments which move through narrow passages in each domain of the sodium channel protein [75, 122 - 124]. The S4 segment in a single domain is a rigid cylinder within the channel with a narrow, hourglass-shaped waist. Positively charged residues of the S4 segment are partially neutralized by interactions with negatively charged residues from the surrounding protein (the S2 and S3 segments). Upon depolarization, the S4 segment moves outward and rotates to place the positively charged residues in more outward positions. Molecular modeling provided a description for the sequence of conformational changes and gating charge movement of the voltage sensor during activation [125, 126]. The S4 segment and its gating charges move through a narrow gating pore that focuses the transmembrane electric field to a distance of approximately 5 Å normal to the membrane (see also [127, 128]). It should be stressed that residues Val109 and Leu112 are the key structures participating in the charge transfer mechanism [129, 130]. These model representations give a structural insight into molecular mechanism of the effective charge transfer in activation gating machinery, thus being the starting point for the further analysis of voltage sensitivity of an ion channel and its modulation due to intermolecular coupling or due to ligand-receptor interactions.

LIMITING SLOPE PROCEDURE

The voltage dependence of conductance of sodium channels is rather steep, which means that the aminoacid sequence of the channel contains a voltage sensor detecting the transmembrane electric potential difference E. This region should bear an electric charge that could be displaced upon a change in the E value, consequently inducing conformational transition of the channel gating machinery from the closed to the open state. It is known that the background intensive parameter thermodynamically coupled to the extensive parameter E is the charge displacement, q, required to open one sodium channel. Though the real value of q is not known with certainty, it seems intuitively to be related to the steepness with which the sodium conductance depends on the membrane potential, and it can be evaluated using a very simple idea introducing the equilibrium constant K_{Na} [58, 62]:

$$K_{Na} = G_{Na}/(G_{Na}^{max} - G_{Na}),\qquad(1.1)$$

where G_{Na}, the sodium conductance, is assumed to be proportional to the number of open sodium channels, while G_{Na}^{max} is the value with all channels open. The difference between G_{Na}^{max} and G_{Na} is evidently proportional to the number of closed sodium channels. Then K_{Na}, the ratio of open to closed channels, is regarded as voltage-dependent equilibrium constant, and $d(\ln K_{Na})/dE$ can be defined as the logarithmic voltage sensitivity of the sodium channel. If the channel gating device were a two-state, open/closed system and the Boltzmann principle could be applied, one would find that [62]:

$$\frac{d\,(\ln K_{Na})}{dE} = \frac{q}{k \cdot T},\qquad(1.2)$$

where E is the membrane potential, k is the Boltzmann constant, and T is the absolute temperature. At potentials so negative that $G_{Na} \ll G_{Na}^{max}$, G_{Na} grows exponentially with E [58]:

$$G_{Na} = const \cdot e^{(q \cdot E)/(k \cdot T)}.\qquad(1.3)$$

At these potentials the peak sodium conductance grows e-fold within 4 mV [62, 131], suggesting by equations (1.1)-(1.3) that q equals to 6 elementary charges. It should be stressed that this estimate of the q value is model-independent. In other words, Alan Hodgkin and Andrew Huxley were the first to evaluate the charge displacement required to open a classical sodium channel using the approach that is quite general and independent on the real physical mechanism of channel gating. Now the family of voltage-dependent sodium channels includes nine representatives [70, 72] which are distinguished from each other by characteristic features of their gating machineries. Herein we consider solely the behavior of the activation gating device of $Na_V1.8$ channels. Our data indicate that modulation of their gating machinery has an important physiological effect: a decrease of the effective charge transfer decreases the excitability of nociceptive neuron membrane and, consequently, results in pain relief in humans. It is demonstrated below how the effective charge transfer in the activation gating device of $Na_V1.8$ channels can be evaluated applying the Hodgkin-Huxley theory.

Inactivation gating machinery of $Na_V1.8$ channels is so slow that it does not distort peak sodium current measurements. This is why application of the logarithmic voltage sensitivity method, also well known as the Almers' limiting slope procedure [58], allows to reliably evaluate q. This approach based on experimental patch-clamp sodium current recordings helps to find novel mechanisms of physiologically adequate modulation of $Na_V1.8$ channels.

The voltage dependence of the sodium conductance $G_{Na}(E)$ in response to constant voltage steps E can be obtained after measuring the maximum (peak) values of the sodium current, $I_{max}(E)$:

$$G_{Na}(E) = I_{max}(E)/(E - E_{Na}), \qquad (1.4)$$

where E_{Na} is the reversal potential of the sodium current. $G_{Na}(E)$ is a monotonic function which approaches its maximum (G_{Na}^{max}) at positive potentials E.

Assuming that the channel gating device is a two-state open/closed system, the Boltzmann principle can be applied at equilibrium:

$$K_{Na} = N_o/N_c \rightarrow G_{Na}(E)/\left(G_{Na}^{max} - G_{Na}(E)\right)_{E \rightarrow -\infty} \rightarrow const \cdot e^{(q \cdot E)/(k \cdot T)}, \qquad (1.5)$$

where k is the Boltzmann constant, T is the absolute temperature, the sodium conductance G_{Na} is considered to be proportional to the number of open sodium channels (N_o), $[G_{Na}^{max} - G_{Na}(E)]$ is proportional to the number of closed channels (N_c). The total charge q of the voltage sensor per one channel can be estimated as the product of the elementary electron charge e_0 times the number of charges being displaced. It is worth noting that any physical measurements of voltage sensitivity of sensory neuron membrane would yield the value equal to the total charge per channel (q) times the fraction of the electric field (ΔE) crossed by the displacing charges, but not the absolute values of any of these factors. This is why the term "effective charge" (Z_{eff}) is introduced.

Z_{eff} can be obtained using the Almers' limiting slope procedure on the basis of equations (1.4)-(1.5):

$$N_o/N_c \rightarrow G_{Na}(E)/\left(G_{Na}^{max} - G_{Na}(E)\right)_{E \rightarrow -\infty} \rightarrow const \cdot e^{(Z_{eff} \cdot e_0 \cdot E)/(k \cdot T)} . \qquad (1.6)$$

Equation (1.6) allows to evaluate Z_{eff} by constructing the voltage dependence of the logarithmic voltage sensitivity function $L(E)$:

$$L(E) = \ln(G_{Na}(E)/(G_{Na}^{max} - G_{Na}(E))). \qquad (1.7)$$

The slope of the asymptote passing through the first points of this function obtained at the most negative values of the membrane potential E makes it possible to calculate the Z_{eff} value, which is linearly proportional to the tangent of the asymptote slope.

Mathematical Modeling of Z_{eff} Evaluation by the Limiting Logarithmic Voltage Sensitivity Method

Using the Hodgkin-Huxley model, $Na_V 1.8$ slow sodium current evoked by constant voltage step E can be presented as

$$I_{Na}(t) = G_{Na}^{max} \cdot m^3(t) \cdot h(t) \cdot (E - E_{Na}), \qquad (1.8)$$

where m and h are the Hodgkin-Huxley variables which describe activation and inactivation gating mechanisms and obey first-order kinetics. Transition of the activation gate from the closed $(1 - m)$ to the open (m) state is governed by differential equation (1.9):

$$\frac{dm}{dt} = \alpha_m \cdot (1 - m) - \beta_m \cdot m , \qquad (1.9)$$

where α_m and β_m are the forward and backward rate constants of activation process. Transition of inactivation system from the closed $(1 - h)$ to the open (h) state is determined by similar equation (1.10):

$$\frac{dh}{dt} = \alpha_h \cdot (1 - h) - \beta_h \cdot h , \qquad (1.10)$$

where α_h and β_h are the inactivation process rate constants. Mathematical modeling allows to investigate characteristic features of interaction between activation and inactivation gating machineries, roles of their gating charges and, finally, excitability of nociceptive neuron membrane.

Experimental voltage dependencies of $Na_V1.8$ channel rate constants (s-slow) were obtained from our patch-clamp experiments:

$$\alpha_{m\_s} = e^{(0.039 \cdot E - 0.56)}, \qquad \beta_{m\_s} = e^{(-0.049 \cdot E - 2.53)}, \qquad (1.11)$$

$$\alpha_{h\_s} = 0.002 \cdot e^{-E/30}, \qquad \beta_{h\_s} = 0.1/(1 + 0.2 \cdot e^{-(E+10)/7}). \qquad (1.12)$$

Functions (1.12) are exactly based on our experimental data describing inactivation process throughout physiological range of E. Functions (1.11) represent a somewhat simplified version of actual experimental curves (see equations (1.18) below). They can be used only in the most negative E range and contain information concerning the effective charge transfer in activation system of $Na_V1.8$ channels. This information can be easily derived applying the Boltzmann principle (see equation (1.17)). Independent verification of the limiting slope procedure can be performed using equations (1.8)-(1.12), which make it possible to calculate the families of I_{Na} currents generated in response to constant voltage steps for different models of channel gating. Then we can simulate the experimental Z_{eff} evaluation procedure on the basis of the Almers' limiting slope method.

m-model

The time dependence of the sodium current triggered by a voltage step of constant value E can be described by the following equation for the single m-particle model:

$$I_{Na}(t) = G_{Na}^{max} \cdot m(t) \cdot (E - E_{Na}). \qquad (1.13)$$

As the first step to simulate the experimental logarithmic voltage sensitivity method, we calculated the family of sodium currents (Fig. **1.1.a**, inset) obtained in response to different E steps in physiological range. Z_{eff} was evaluated using the $L(E)$ function (1.7). It was equal to 2.25 elementary charge units (Fig. **1.1.a**). This result is in satisfactory accordance with the value estimated for voltage sensor gating in holo-proton conductive H_V channels [132]. Their S1-S4 transmembrane segments are similar to the corresponding parts of sodium channels. This "voltage sensor domain" is located within the main H_V channel subunit. Ile262 and Asn264 were demonstrated to be more accessible to the cytoplasmic solution in the closed state, as compared to the open state of the channel [132]. These observations are consistent with the crystallographic data reported for the S4 segments of K_v channels, and it was further estimated that an equivalent of 2–3 gating charges associated with S4 might move in a single H_v subunit [132, 133].

m^3-model

This model correctly explains the voltage-clamp data for activation gating systems of fast ($Na_v1.1$) and slow ($Na_v1.8$) sodium channels. Fig. (**1.1.b**, inset) displays the time dependence of slow sodium currents calculated by equation (1.14):

$$I_{Na}(t) = G_{Na}^{max} \cdot m^3(t) \cdot (E - E_{Na}).$$

(1.14)

Using the peak values of sodium currents, the logarithmic voltage sensitivity function $L(E)$ was plotted. The tangent of its slope yielded $Z_{eff} = 6.75$ (Fig. **1.1.b**). This theoretical value obtained in physiological range of E is not affected by inactivation process that might decrease the amplitude of the peak sodium current. Further calculations will demonstrate the limitations of application of the Almers' procedure strongly depending on the kinetics of inactivation process.

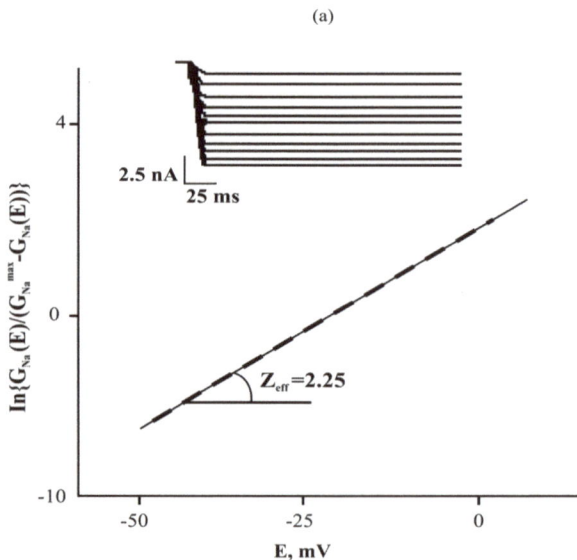

Fig. 1.1 contd.....

(b)

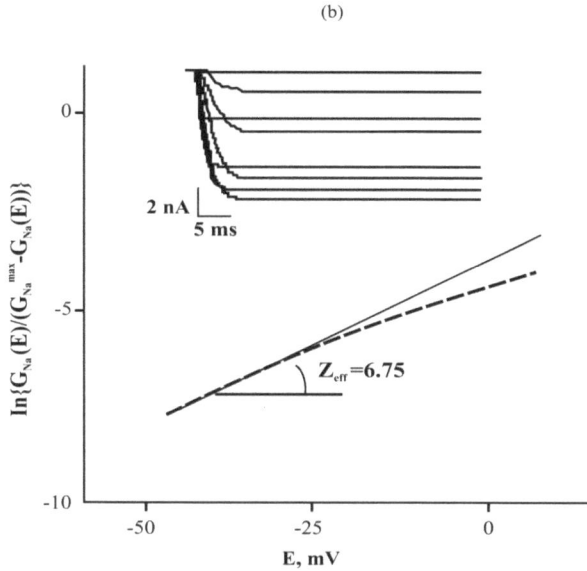

Fig. (1.1). The Almers' procedure for evaluation of the effective charge using the Hodgkin-Huxley theory for $Na_V1.8$ channel activation gating system.

a – The family of sodium currents was calculated using the m-model (inset). Their peak (amplitude) values were used to simulate the logarithmic voltage sensitivity function $L(E)$, dashed straight line. Its asymptote (solid line) plotted through points obtained at very negative E fully coincides with $L(E)$ function (the simplest linear condition for "elementary" m-model).

b – The family of sodium currents was calculated using the m^3-model (inset). Their peak (amplitude) values were used to simulate the logarithmic voltage sensitivity function $L(E)$, dashed line. It is clearly seen that its asymptote (solid line) plotted through points obtained only at negative E coincides with the $L(E)$ function (nonlinear condition for the m^3-model).

m^3h_s-model

Our patch-clamp data (equations (1.11)-(1.12) were used to calculate $Na_V1.8$ slow sodium currents. Their kinetic behavior is also governed by inactivation system (functions (1.8)). Fig. (**1.2.a**, inset) presents the results of simulation of slow sodium currents family in voltage-clamp conditions. The peak value of each current can be easily measured, which allows to construct the $L(E)$ function (equation (1.7)). Application of the Almers' limiting slope procedure gives the Z_{eff} value equal to 6.2 (Fig. **1.2.a**), and it is close to the result obtained using the m^3-model. Inactivation process affects the peak value of the sodium current. Apparently, the faster kinetics of inactivation system is the bigger the discrepancy should be. That is why our next step involved simulation of the peak currents of a "chimeric channel", the activation gate of which is slow ($Na_V1.8$), while the inactivation gate is fast ($Na_V1.1$, equation (1.15)).

(a)

(b)

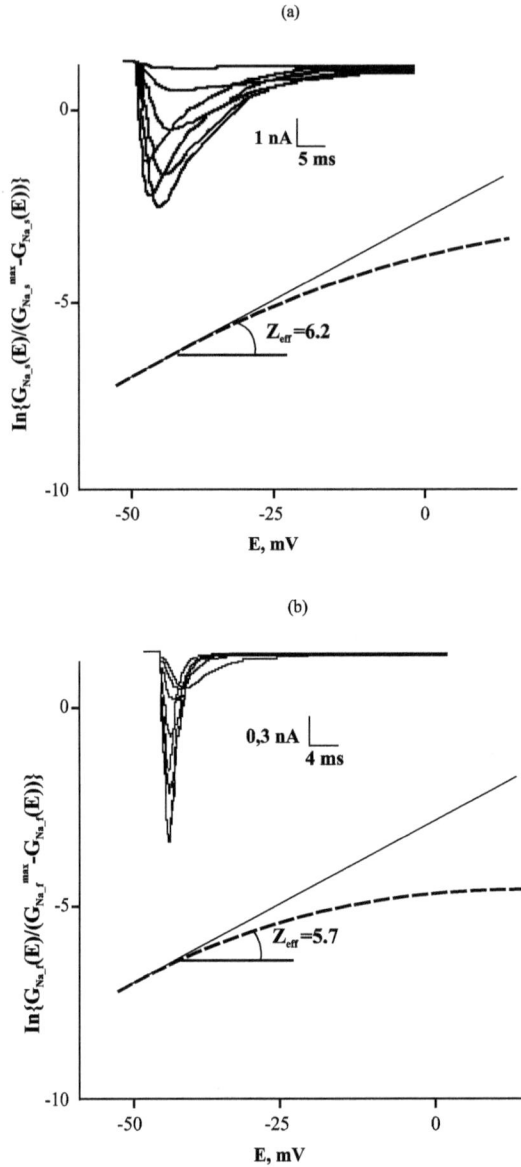

Fig. (1.2). Accuracy of results obtained by the Almers' method depends on kinetics of inactivation process.
a – The family of sodium currents was calculated using the m^3h_s-model (inset). The peak value of each simulated $Na_V1.8$ current was measured and the $L(E)$ function was constructed. Application of the Almers' limiting slope procedure for the model which takes inactivation gating process into account gives Z_{eff} value close to the result obtained using the m^3-model.
b – The family of chimeric channel sodium currents was calculated using the m^3h_f-model (inset). The process of their inactivation was accelerated as compared to $Na_V1.8$ channels. The Z_{eff} value derived by the limiting logarithmic voltage sensitivity method is one elementary charge less than that obtained using the m^3-model. Therefore, application of the Almers' method to determine the effective charge of fast sodium channels should be implemented with caution.

m³h_f-model

Fig. (**1.2.b**) illustrates the family of sodium currents (inset) calculated using experimentally obtained "f-fast" inactivation rate constants [134]:

$$\alpha_{h\_f} = 0.012 \cdot e^{-(E+43)/10}, \quad \beta_{h\_f} = 1.32/(1+0.2 \cdot e^{-(E+10)/7}). \quad (1.15)$$

In this case the Z_{eff} value derived by the limiting logarithmic voltage sensitivity method equals to 5.7 only (Fig. **1.2.b**). Acceleration of inactivation gating process results in the "loss" of one elementary charge as compared to the m^3-model.

Calculations based on experimental voltage dependencies of fast (equations (1.15)) inactivation rate constants demonstrate a pronounced decrease in Z_{eff} of the chimeric channel. The deviation from "ideal" q value is much lower for slow inactivation process which is an inherent characteristic of slow $Na_V1.8$ channels. This is why it can be concluded that the Almers' limiting slope procedure is applicable for evaluation of the effective charge transfer in these channels with a satisfactory accuracy.

Thermodynamic approach based on the Boltzmann principle makes it possible to evaluate the gating charge analyzing experimental voltage-clamp data of rate constants' voltage dependencies, which should be presented as exponential functions [135]. Inactivation system of fast sodium channels of nodal membrane obeys second-order properties [136, 137] and transfers part of its charge between two closed states and the other part between the open and closed states. Asymmetrical gate charge transfer in this three-state system is the cause of spike frequency adaptation of nerve fiber membrane [69], and it might explain the peculiarities of phasic and tonic responses of sensory and motor nerve fibers [69]. Here we apply the same approach, the crucial point of which is the exponential voltage dependence of each Hodgkin-Huxley variable that should be presented as

$$\alpha_m = e^{(a_m \cdot E + b_m)}, \quad \beta_m = e^{(c_m \cdot E + d_m)}, \quad (1.16)$$

where a_m, b_m, c_m, d_m are constants. At equilibrium $dm/dt = 0$, thus applying equation (1.9) and the Boltzmann principle one can obtain

$$N_o/N_c = m/(1-m) = \alpha_m/\beta_m = e^{Z_{eff} \cdot e_0 \cdot (E-E')/k \cdot T}, \quad (1.17)$$

where E' is the potential at which the channels are equally distributed between the open and closed states. Using our experimental data for α_m and β_m functions, equation (1.11), Z_{eff} calculated from equation (1.17) is 2.25 and exactly coincides with the value obtained by application of the logarithmic voltage sensitivity function for the m-model (Fig. **1.2.a**), which is a trivial result for the case of this linear model. When nonlinear characteristics are involved to describe the gating machinery behavior (the m^3-model), the results are not so obvious.

Fig. (**1.3**) illustrates m- and m^3-model gating charge transfer behavior of the activation device. If the activation device obeys first-order kinetics (Fig **1.3**, top), only one gating particle transfers the charge $q/3$ equal to 2.25. The total charge q, which should be equal to 6.75 (a three times greater value), can be transferred by a more complicated gating machinery consisting of three independent particles (Fig. **1.3**, bottom). The latter model accounts for the experimental voltage-clamp data. Application of the Almers' limiting logarithmic voltage sensitivity method yields the same value: $Z_{eff} = 6.75$ (Fig. **1.2.b**). The $L(E)$ function deviates from the limiting asymptote at depolarizing E due to nonlinear characteristics of m^3-gating. This effect is much more important when inactivation process is involved in sodium channel opening. It interferes with the activation machinery and slightly decreases Z_{eff} to 6.2 (Fig. **1.2.a**). But this loss of half of elementary charge is not critical for applicability of the Almers' limiting slope procedure. It is clear that "ideal" q value can never be registered in physiologically adequate conditions. All our measurements are based on comparison of the control data with the data obtained at the same conditions after application of an agent under investigation. This comparison makes it possible to elucidate the mechanisms of interaction between the gating machinery of sodium channels and neighboring protein molecules, which together control intracellular signaling and physiological encoding processes (see below).

Analyzing experimental data on voltage-dependent sodium channels obtained by different methods, it becomes evident that molecular mechanisms of gating charge transfer are far from deep understanding. Nevertheless, it is demonstrated that each sodium channel transfers a constant amount of gating charge per channel, which varies from 4 to 8 [62, 89, 138 - 140], and in certain cases up to 12 elementary charges [141]. This scatter of experimental data for sodium channels can be explained by the facts that the channel expression was assumed to be too small and the sodium channel kinetics too fast for the measurement of gating currents by conventional voltage-clamp recording techniques [142]. However, upon generation of trains of nerve impulses in adequate physiological conditions the effective charge transfer of sodium channels should fall into a substantially more narrow range (roughly from 3 to 7 elementary charge units).

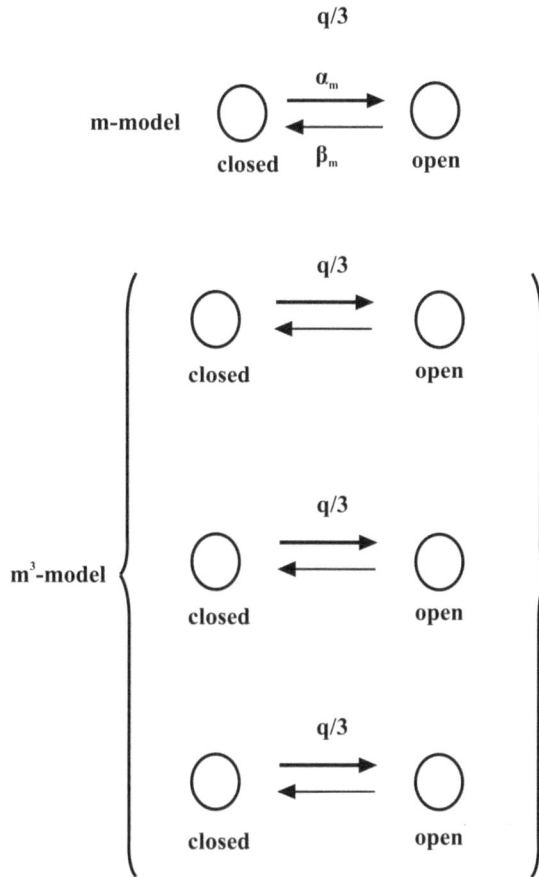

Fig. (1.3). Scheme illustrating gating charge transfer in the *m*- and *m³*-model.
The activation device obeys first-order kinetics (top). The only one gating particle transfers the elementary charge *q/3*.
The total charge *q*, which should be a three times greater value, is transferred by a more complicated gating machinery consisting of three independent particles (bottom).

Previously we have predicted that the sodium channel inactivation device also transfers the gating charge, the values of which differ between sensory and motor nerve fibers [69]. The presence of voltage sensor in inactivation gating machinery of the squid giant axon membrane was also postulated [143]. The charge equal to 1.2 elementary units was found to be displaced during inactivation [144]. These studies suggested that the voltage sensor for inactivation process was most likely located in the segment S4 of domain 4 (D4) of fast sodium channel. By investigating ionic currents of mutant channels, S4D4 was demonstrated to control inactivation, probably being directly involved in the process [145 - 149].

Detailed analysis of fine tuning of gating machinery has been provided mostly for potassium channels expressed in *Xenopus* oocytes [150 - 153]. The mechanism by which $K_V1.2$ channels sense changes in cell membrane voltage was investigated by crystallographic methods [154]. The voltage sensor was suggested to perform mechanical work on the pore through the S4-S5 linker helices, positioned as to constrict or dilate the S6 inner helices of the pore. In the open conformation, two of the four conserved Arg residues on S4 are located on a lipid-facing surface and the other two are buried within the voltage sensor. It was also predicted that with the opening of *Shaker* potassium channel, the net equivalent of 12 to 14 positive elementary charges was transferred across the membrane electric field from inside to outside, and this charge was mostly carried by four S4 Arg residues on each of four identical channel subunits [155 - 157].

Another potassium channel $K_V7.1$ (KCNQ1–KCNE1 complex) plays different physiological roles in heart and epithelial tissues demonstrating its flexibility in terms of the mechanism of channel gating. The KCNQ1 channel was found to be regulated by several endogenous factors, in particular, ATP, PIP2, and PKA. ATP promotes KCNQ1–KCNE1 channel opening due to the weak S4-to-gate coupling [133, 158]. This proposed ATP binding site differs from classical ATP binding sites, *e.g.*, those in P2X receptors, and it is probably located close to S6 gate [158]. The acidic phospholipid PIP2 also promotes KCNQ1 channel opening. PIP2 is suggested to bind to a cluster of basic residues in the intracellular S2–S3 and S4–S5 loops of KCNQ1 [159]. The putative PIP2 binding site bridging the voltage-sensor domain and the pore domain is proposed to enable the PIP2 molecule to communicate conformational changes in the voltage-sensor domain to the pore domain, and *vice versa* [133, 160, 161]. Another modulator of $K_V7.1$ channels, PKA, phosphorylates the N-terminus of KCNQ1 channels in the presence of the anchoring protein yotiao, which binds to the C-terminus of KCNQ1, thus affecting the S4 movement, the gate, or the coupling between S4 and the gate [133, 162]. These findings show that gating devices of voltage-dependent potassium channels are under control of endogenous substances involved in intracellular signaling mechanisms, which are physiologically adequately activated by membrane receptors. There is a great number of receptor-regulated ionic channels, but just a few of them change their gating charge as a result of interaction of an endogenous attacking molecule with a membrane receptor coupled to the channel [163, 164]. Membrane receptors are usually coupled to G proteins that amplify and transduce the signals triggered by attacking molecules in two directions: to the cellular genome and to ion channels.

Neurotransmitters and endogenous peptides were firstly demonstrated to effectively modulate voltage-dependent calcium channels through activation of membrane receptors and coupled G proteins [165 - 167]. Electrical excitability of rat hippocampal neurons was shown to be regulated by sodium channels coupled to G proteins [168]. There are indications that the gating charge machinery is involved in coupling between calcium channel and G protein [169 - 171]. The functioning of ion channels was thus demonstrated to be modulated not only by activation of a membrane receptor but also by direct action on a transducer molecule (in this case, G protein) involved in signal transduction from the receptor to the channel. Using the limiting slope method, the effective charge of Ca(V)2.2 channels was shown to diminish from 6.3 to 4 elementary units after application of GTPγS, a G protein activator. Increased concentration of noradrenaline induced a decrease of the effective charge transfer which was under G protein control [171]. Application of the limiting slope procedure to neuronal and cardiac calcium channels gave the charge per channel value equal to 8.6 [172].

Investigation of gating events at the single channel level based on the analysis of fluctuations of the ensemble gating currents has yielded information on an elementary event of effective charge transfer required to open a single channel [173 - 175]. In particular, effective charge transferred during an elementary event in mammalian fast sodium channels is 2.3 [144]. Application of this technique to *Shaker* potassium channel led to a similar result: the elementary charge per event was estimated to be 2.4 for large depolarizations [174, 176, 177].

These values are in a very good accordance with our theoretical and experimental data obtained on $Na_V1.8$ channels. Application of the Almers' limiting slope procedure for the simplest gating device in the framework of the first-order (linear) m-model demonstrates that the charge transferred per elementary event during $Na_V1.8$ channel activation is 2.25 (Fig. **1.1.a**), which is very close to the corresponding values for fast sodium channels and *Shaker* potassium channels. Our independent calculation of effective charge obtained from experimental rate constants using the Boltzmann principle yielded a value of 2.3 (equation (1.17)), while three activation gating m-particles (the m^3–model) will transfer 6.9 elementary charges.

It is tempting to suppose that activation gating machinery of slow sodium channels plays a key role in control of sensory neuron excitability. Effective charge of $Na_V1.8$ channel activation gating device can be expected to vary from 3 to 7 elementary units. If the membrane potential E falls outside the physiologically adequate diapason, the effective charge range expands to 2–8 elementary units.

It will be shown below that our experimental approach based on the Almers' limiting slope method makes it possible to quantitatively analyze modulation of nociceptive membrane excitability by measuring Z_{eff} and obtain dose-response functions after addition of a number of agents applying for the role of an analgesic. The agents which decrease the effective charge transfer of $Na_V1.8$ channel activation gating system should consequently inhibit impulse firing of nociceptive neuron. To demonstrate that fine tuning of nociceptive membrane impulse firing can be controlled exclusively by effective charge modulation while other channel characteristics remain unchanged, one should perform mathematical analysis of the Hodgkin-Huxley model of nociceptive neuron membrane. This approach helps us to reveal novel nociceptive membrane excitation control mechanisms which might be responsible for pain relief in humans.

PROBABLE ROLE OF EFFECTIVE CHARGE TRANSFER IN NOCICEPTIVE INFORMATION CODING

Slow (tetrodotoxin-resistant, TTX-R) sodium channels were discovered using the Kostyuk's method [95] and were later shown to be responsible for nociception in mammals [96, 97], which opened a radically new prospect in development of analgesics [178]. $Na_V1.8$ channels are thus far found to be expressed by the gene SCN10A in neurons of the peripheral nervous system [179, 180], in cardiac tissue [181], in cerebellar Purkinje neurons [182, 183] and enteric neurons [184]. These channels control the spike frequency generation of nociceptive membrane and also finely modulate this firing due to peculiar structure of their gating devices. We predict that $Na_V1.8$ channel activation gating machinery is responsible for control of nociceptive signals. Gating charge transfer can be effectively modulated by a number of endogenous and exogenous substances, which enables fine tuning of nociceptive membrane impulse firing. This mechanism of primary sensory coding strongly depends on $Na_V1.8$ channel gating characteristics.

In currently accepted theories of nociception, pain relief is accounted for by a decrease of $Na_V1.8$ channels density [185 - 188]. We predict that nociceptive membrane impulse firing is controlled by gating machineries of these channels, primarily by their activation gating devices. $Na_V1.8$ channel effective charge transfer is modulated by coupled neighboring membrane and submembrane proteins. Investigating the relationship between $Na_V1.8$ channels and coupled protein molecules, one can find novel signaling cascades, triggering which one might specifically modulate the activation gating device of these channels and therefore control the nociceptive neuron firing frequency. Decrease of gating charge transfer is an additional mechanism to achieve pain relief in mammals.

Our experimental data obtained by the current clamp method demonstrate that sensory neuron membrane incorporating both fast and slow sodium channels generates single action potential (AP), the shape of which is only slightly modified after fast sodium channels have been switched off (Fig. **1.4**). This fact indicates that $Na_V1.8$ channels play a key role in generation of sensory neuron impulse firing, comparable to the role of classical $Na_V1.1$ channels. Sensory membrane with fast sodium channels being blocked by TTX and thus including only one type of sodium channels, $Na_V1.8$ channels, discharges trains of nerve impulses (Fig. **1.4.b**). The characteristic feature of this train is the increase of interspike intervals in time. This adaptational phenomenon is accomplished by the decrease of under- and overshoots of AP amplitudes which indicates involvement of inactivation gating system in control of impulse firing [69].

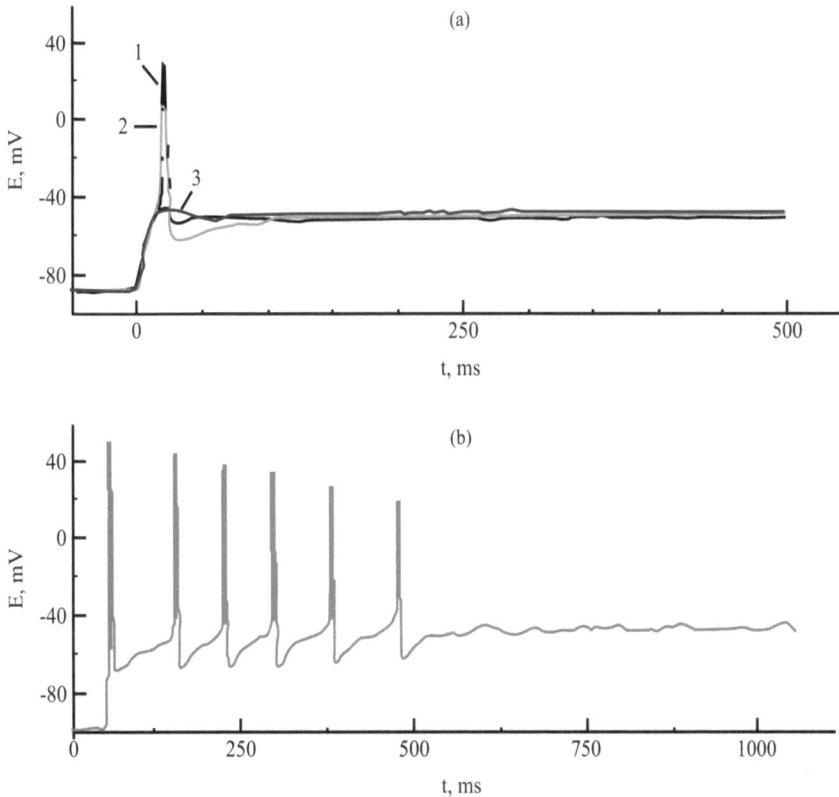

Fig. (1.4). Current clamp recordings of sensory neuron firing.
a – AP generation in response to constant current step. 1 – (black line) response with two populations of sodium currents involved in impulse firing. 2 – (gray line) TTX-sensitive current is switched off, $Na_V1.8$ channels are involved in inward current generation. 3 – (dark gray line) all sodium currents are blocked by tetracaine, $I_{stim} = 150$ pA.
b – Impulse firing generated by TTX-resistant sodium channels after TTX application, $I_{stim} = 195$ pA.

Results of our patch-clamp whole-cell experiments on sensory neuron membrane make it possible to quantitatively describe all characteristics of $Na_V1.8$ channels in terms of the Hodgkin-Huxley model (Fig. **1.5**). In particular, this approach helps to clarify the physiological role of gating charge transfer in nociceptive signal coding. Our assumption that nociceptive signals are specifically tuned exclusively by modulation of activation gating charge transfer can be verified only by mathematical analysis of such a theoretical model of nociceptive membrane, in which the only variable is Z_{eff} and all other $Na_V1.8$ channel characteristics remain unchanged. It should be stressed that these conditions can hardly be achieved in experiments, which is why our results regarding primary sensory coding obtained using mathematical modeling form the background of the further investigation strategy of pharmacological control over the nociceptive system.

Nociceptive neuron membrane impulse firing can be simulated by constructing the Hodgkin-Huxley model of six nonlinear differential equations:

$$C_m \cdot \frac{dE}{dt} = I_m - I_{Na\_s} - I_{Na\_f} - I_K - I_L \,,$$

$$\frac{dm_{Na\_s}(E,t)}{dt} = \propto_{mNa\_s}(E) \cdot (1 - m_{Na\_s}(E,t)) - \beta_{mNa\_s}(E) \cdot m_{Na\_s}(E,t) \,,$$

$$\frac{dh_{Na\_s}(E,t)}{dt} = \propto_{hNa\_s}(E) \cdot \left(1 - h_{Na\_s}(E,t)\right) - \beta_{hNa\_s}(E) \cdot h_{Na\_s}(E,t) \,,$$

$$\frac{dh_{Na\_f}(E,t)}{dt} = \propto_{hNa\_f}(E) \cdot \left(1 - h_{Na\_f}(E,t)\right) - \beta_{hNa\_f}(E) \cdot h_{Na\_f}(E,t) \,,$$

$$\frac{dm_{Na\_f}(E,t)}{dt} = \propto_{mNa\_f}(E) \cdot (1 - m_{Na\_f}(E,t)) - \beta_{mNa\_f}(E) \cdot m_{Na\_f}(E,t) \,,$$

$$\frac{dn(E,t)}{dt} = \propto_{n}(E) \cdot \left(1 - n(E,t)\right) - \beta_{n}(E) \cdot n(E,t) \,.$$

The first master equation here is consistent with the Kirchhoff's law for electrical circuit, where the total current I_m branches off into the displacement capacitive current (C_m is the membrane capacitance) and currents carried by the corresponding ions through ion channels. I_{Na_s}, I_{Na_f}, I_K, I_L are the slow sodium current, the fast sodium current, the potassium current, and the leakage current, respectively. The Hodgkin-Huxley variables m, n, and h vary in the range from 0 to 1. The following parameters were accepted in the model: maximum value of fast sodium channel conductivity $G_{Na\_f}^{max}$ = 50 nS, maximum value of potassium channel conductivity G_K^{max} = 40 nS, membrane capacitance C_m = 20 pF, sodium current reversal potential E_{Na} = 55 mV, potassium current reversal potential E_K = -85 mV, resting potential E_r = -70 mV, leakage current reversal potential E_L = -70 mV, leakage current conductivity G_L = 5 nS.

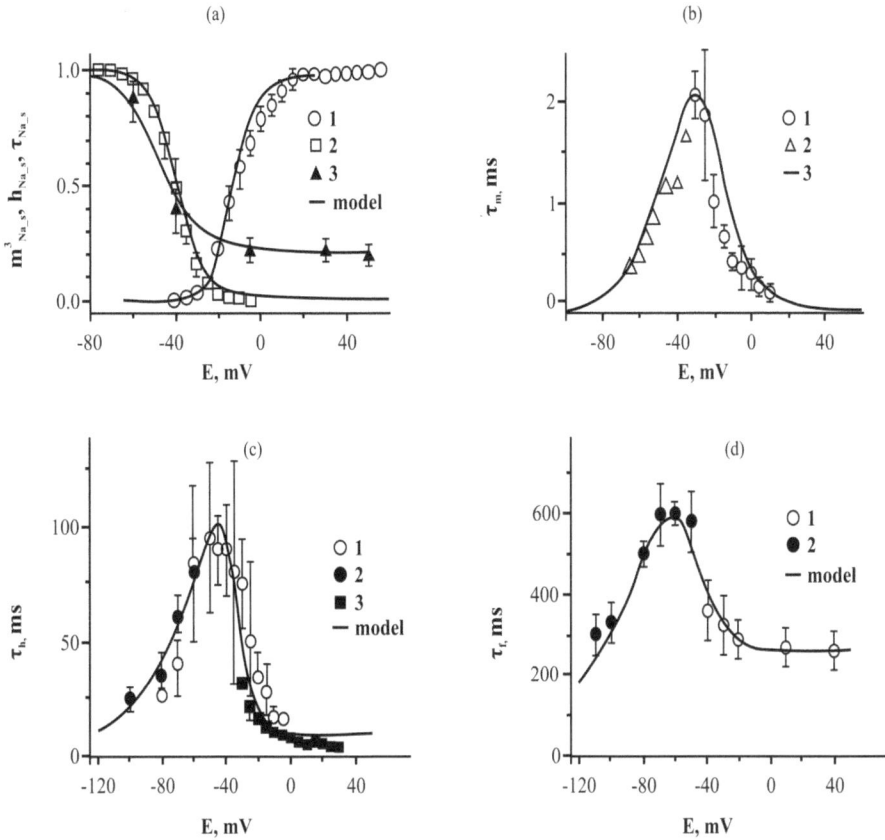

Fig. (1.5). Gating characteristics of $Na_V1.8$ channels.

a – Hodgkin-Huxley steady-state gating variables of slow sodium channels. Steady-state activation and inactivation voltage dependencies are obtained experimentally and fitted by solid curves calculated using voltage dependencies of the corresponding rate constants. 1, 2, 3 – experimental steady-state values of activation $m_{Na\_s}^3(E)$, inactivation $h_{Na\_s}(E)$, and slow inactivation functions $r_{Na\_s}(E)$, respectively.

b – Activation gate time constant as a function of the testing potential $\tau_m(E)$. 1 – $\tau_m(E)$ obtained during membrane depolarization to different values of square test pulses $E > -40$ mV. 2 – $\tau_m(E)$ obtained during investigation of deactivation process using the double pulse method where membrane was repolarized from 0 mV (duration of this conditioning pulse was equal to 10 ms), test pulses values $E < -30$ mV.

c – Voltage dependence of fast inactivation time constant $\tau_h(E)$. 1 – recovery from inactivation, double pulse experiment, $-120 < E < -70$ mV; 2 – development of inactivation, double pulse experiment, $-80 < E < 0$ mV; 3 – development of inactivation, $-40 < E < 40$ mV.

d – Voltage dependence of slow inactivation time constant $\tau_r(E)$. 1 – single pulse experiments of development of inactivation, $E > -40$ mV; 2 – double pulse experiments of recovery from inactivation, $E < -40$ mV.

Solid lines represent the results of calculations based on equations describing voltage dependencies of the corresponding rate constants (model).

Rate constant functions required to perform the calculations were obtained from our patch-clamp data presented above (Fig. **1.5**) and from findings of other authors [134, 189]. The patch-clamp technique allows obtaining the voltage dependencies of $Na_V1.8$ channel activation rate constants in a wide range of E

(Fig. **1.6**, solid lines). It is well known that all experimental voltage dependencies cannot be adequately described using single exponential functions if activation gating system obeys first-order kinetics. Nevertheless, we have successfully approximated β_{mNa_s} with a single exponential voltage dependence. The other activation rate constant α_{mNa_s} may be considered as a single exponential function only at negative values of the membrane potential (Fig. **1.6**, dashed lines). The effective charge value calculated using the Boltzmann distribution on the basis of these voltage dependencies (equation (1.12)) was equal to 6.75 for three independent activation particles obeying sodium current m^3-gating. These exponential functions can be as well utilized to calculate the peak sodium currents describing also the behavior of inactivation gate and thus verify applicability of the Almers' limiting slope method to $Na_V1.8$ channels (Fig. **1.2.a**). The "control" Z_{eff} value in this case was shown to be equal to 6.2 (the m^3h_s-model). Almost the same control value of Z_{eff} is obtained using the Almers' method if experimental (non-exponential) α_{mNa_s} and β_{mNa_s} functions that adequately describe activation gating machinery in a wide interval of E (Fig. **1.6**, solid lines) are taken into consideration when calculating the peak sodium currents (equations (1.18)):

$$\alpha_{mNa\_s} = \frac{0{,}05 \cdot \left(1 + e^{(E+44)/8}\right)}{1 + e^{(E-7)/10}} \quad, \quad \beta_{mNa\_s} = 0{,}05 \cdot \left(1 + e^{-(E+6)/16}\right). \tag{1.18}$$

(a)

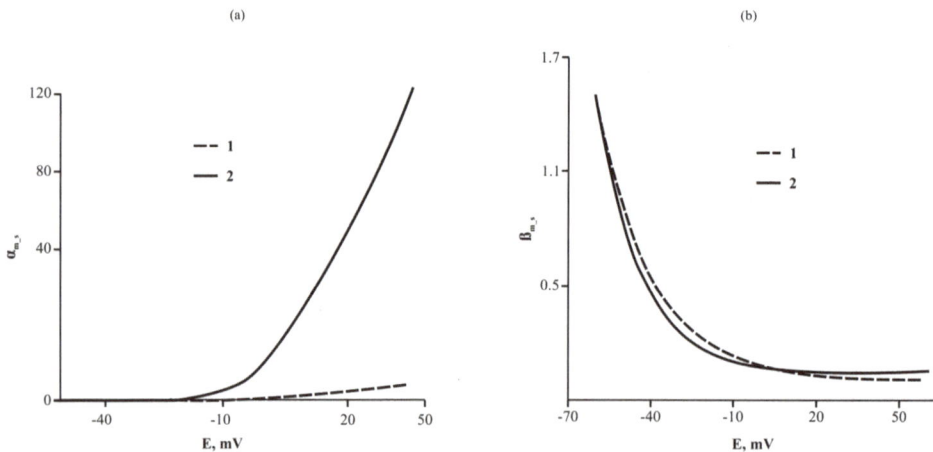

(b)

Fig. (1.6). $Na_V1.8$ channels activation rate constants' voltage dependencies.
Solid lines are obtained from experimental data presented in Figs. (**5.a** and **5.b**) (equations (1.18)). Dashed lines represent results of approximation of these functions at negative E (equations (1.11))

These functions were regarded as the control data (Fig. **1.6**, solid lines). We have further modified them to give an account of the decrease of the effective charge value to 4.5 (equations (1.19)), which corresponds to the effect induced by extracellular application of $Na_V1.8$ channel-modulating agents, in particular, comenic acid (see below):

$$\alpha_{mNa\_s} = \frac{0{,}05 \cdot (1 + e^{(E+40)/9})}{1 + e^{(E-11)/25}}, \quad \beta_{mNa\_s} = 0{,}05 \cdot \left(1 + e^{-(E+2)/20}\right). \qquad (1.19)$$

Functions (1.19) simulate a strong decrease in steepness of the voltage dependence of sodium steady-state activation gating process as compared to functions (1.18), which is accompanied by a weak shift of the activation function to the left along the E axis at very negative E values. In other words, application of functions (1.19) results in a pronounced decrease of $Na_V1.8$ channel voltage sensitivity which, in turn, controls the frequency of nociceptive firing. Nociceptive neuron excitability is also controlled by the voltage dependence of the steady-state activation function, which means that the negative shift of this function along the E axis dramatically decreases the voltage threshold of excitable membrane repetitive firing [69]. This mechanism is shown to enhance the ease of sensory neuron excitation following slow sodium channel phosphorylation by PKA [190]. That is why the valuable effect that might decrease the nociceptive neuron excitability is the positive shift of the steady-state activation voltage dependence along the voltage axis.

Using the Hodgkin-Huxley model, the effect of this positive shift on nociceptive membrane impulse firing can be simulated independently from investigation of Z_{eff} influence on trains of APs. It might be *a priori* predicted that each of these mechanisms could form the basis for antinociceptive reaction in mammals controlled by $Na_V1.8$ channels. Positive voltage shift of the activation function was simulated in theoretical conditions when all parameters, including Z_{eff}, were kept at their corresponding control values and experimental rate constants were shifted 4 mV to the right along the E axis (equations (1.20)):

$$\alpha_{mNa\_s} = \frac{0{,}05 \cdot (1 + e^{(E+40)/8})}{1 + e^{(E-11)/10}}, \quad \beta_{mNa\_s} = 0{,}05 \cdot \left(1 + e^{-(E+2)/16}\right). \qquad (1.20)$$

The results of simulation of "control" firing are presented in Fig. (**1.7**), demonstrating that excitable membrane of nociceptive neuron increases its impulse activity in response to increase of the strength of constant stimulating current. Rhythmical firing occurs when I_{stim} is equal to 120 pA. Its lower value (80 pA) results in generation of just a single AP.

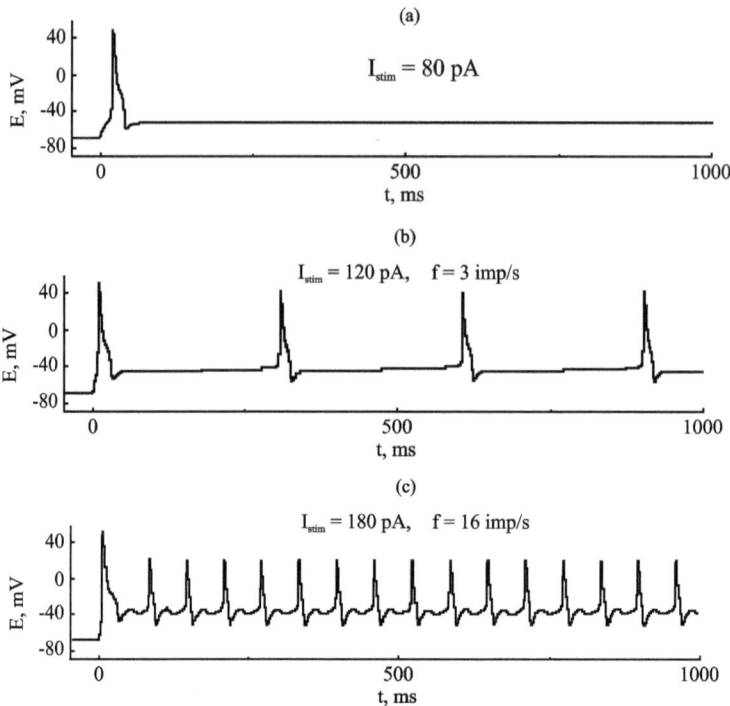

Fig. (1.7). Trains of nerve impulses simulating responses of nociceptive neuron membrane calculated at maintained depolarizing currents (control data).

Fig. (**1.8**) summarizes the effects of positive voltage shift and Z_{eff} decrease on impulse firing of theoretical membrane, which result in a pronounced decrease of impulse frequency as compared to the control data. It can be predicted that both effects are very significant for inhibition of additional nociceptive neuron excitation, which is manifested in the "high-frequency" component of impulse firing.

Generalized results of calculations of repetitive response frequency dependence on the strength of applied maintained current during changes in activation gating system characteristics are presented in the picture of frequency coding performed by nociceptive neuron membrane (Fig. **1.9**). This picture has a very important distinctive feature: Z_{eff} decrease results in a remarkable inhibiting effect. It is clearly seen that the firing frequency never exceeds the threshold level in this case (Fig. **1.9**, solid line), which means that excitable membrane of nociceptive neuron generates exclusively low-frequency responses below the threshold level. The high-frequency component that should transfer nociceptive information to CNS is completely switched off. Remaining low-frequency trains coding thermo-, mechano-, and chemoreceptor information still can be transferred to CNS.

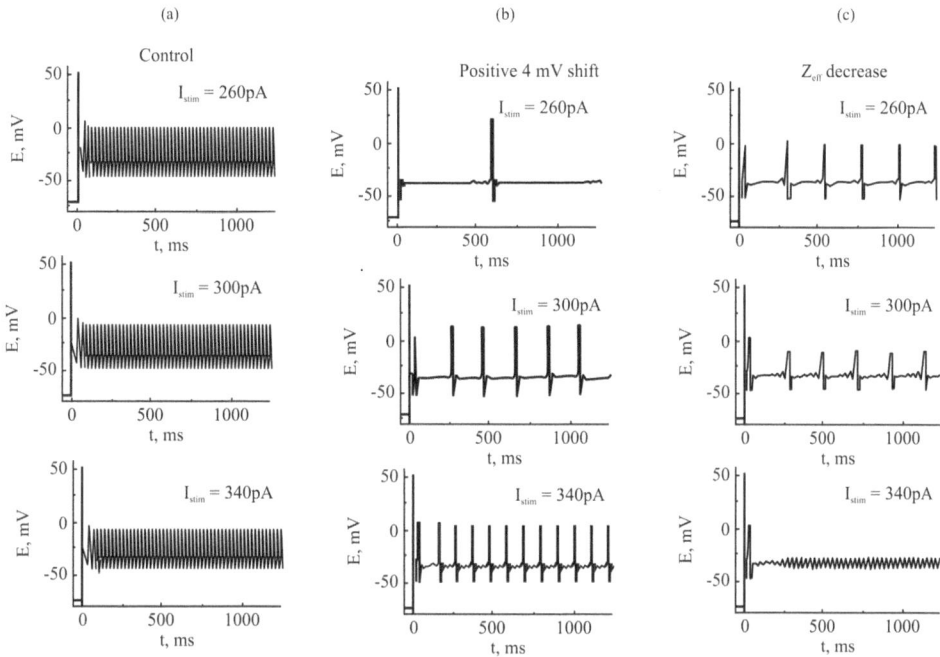

Fig. (1.8). Trains of nerve impulses simulating effects of positive shift and Z_{eff} decrease compared with control responses of nociceptive membrane.
a – Control calculation of membrane firing (equations (1.18))
b – Impulse activity when the steady-state activation function is 4 mV positively shifted (equations (1.20))
c – Impulse activity when Z_{eff} is decreased (equations (1.19)).

On the contrary, positive shift of the activation function along the E axis does not totally switch off the high-frequency nociceptive component which appears again upon increasing the strength of the stimulating current. It means that the working range of nociceptive reaction is also shifted to the right along the abscissa. The firing frequency may exceed a hypothetic nociceptive threshold (Fig. **1.9**, horizontal line) upon increase of the stimulus strength, which indicates that nociceptive signals should be transferred to CNS.

That is why modulation of effective charge transfer has a principal advantage in comparison with other mechanisms of activation gating inhibition. Z_{eff} decrease results in specific inhibition of the high-frequency component of impulse firing responsible for noxious signaling, which forms the basis of our approach to development of novel analgesics specifically modulating effective charge transfer. These substances should inhibit chronic pain without influencing sensory pathways for other modalities transferred by low-frequency discharges.

Our calculations also demonstrate that nociceptive neuron membrane firing is under control of additional slow inactivation process of $Na_V1.8$ channels referred to as r_{Na_s} in Fig. (**1.5**). If $Na_V1.8$ channel slow inactivation process is taken into account, the equation describing the sodium current should be modified as follows:

$$I_{Na\_s} = G_{Na\_s} \cdot m^3_{Na\_s}(E) \cdot h_{Na\_s}(E) \cdot r_{Na\_s}(E) \cdot (E - E_{Na}) \cdot$$

The forward and backward slow inactivation rate constants are:

$$\alpha_{rNa\_s} = 0{,}0001 \cdot e^{-(E+8)/29} + 0{,}0008$$
$$\beta_{rNa\_s} = 0{,}0003/(1 + 0{,}24 \cdot e^{-(E+23)/10}$$

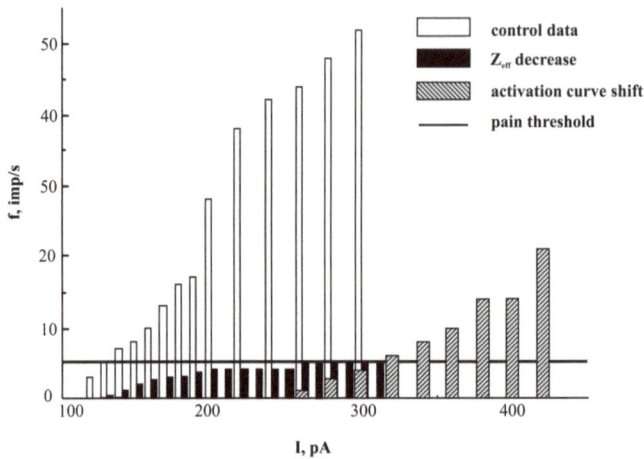

Fig. (1.9). Stimulus-response dependencies of theoretical nociceptive neuron membrane. White columns – responses of control nociceptive neuron; dashed columns – impulse activity is determined by positive shift of the steady-state activation function; black columns – impulse activity is determined by Z_{eff} decrease.
Horizontal line represents a hypothetic "high-frequency" threshold of pain sensation equal to 5 imp/s.

Repetitive firing frequency in this case also depends on the strength of the stimulating current (Fig. **1.10**). Involvement of additional slow inactivation process causes a peculiar change in the repetitive activity which results in a pronounced spike frequency adaptation at moderate levels of the stimulating current (Fig. **1.10.b**). The characteristic feature of human $Na_V1.8$ channels, as opposed to rodent $Na_V1.8$ channels, is a slower inactivation gating process in the former [191]. Nevertheless, at higher values of stimulating currents the discharge frequency exceeds the noxious threshold (Fig. **1.10.c**), which means that slow inactivation process might play an important physiological role in nociceptive signal coding, being though less effective than the decrease of activation gating charge that totally switches off only nociceptive signals.

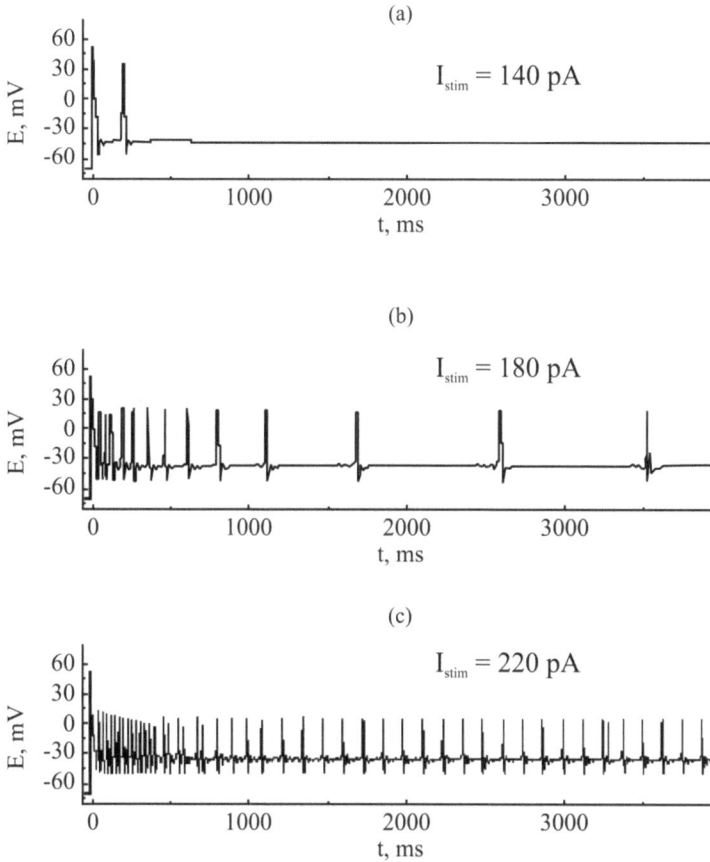

Fig. (1.10). Simulated generation of repetitive responses of nociceptive membrane taking into account slow inactivation gating at different values of the stimulating current.

These high-frequency signals can also be switched off by decreasing the density of functioning $Na_V1.8$ channels (Fig. **1.11**), which is how the majority of pharmaceutical sodium channel modulators produce the antinociceptive reaction or neural dysfunction correction (*e.g.*, [192, 193]). To estimate probable effectiveness of this mechanism with respect to pain relief we have evaluated the amplitudes of stimulating currents required to produce impulse firing with the frequency exceeding the "pain sensation" threshold of 5 imp/s at different values of the maximum conductivity $G_{Na\_s}^{max}$ (Fig. **1.12**) and demonstrated that a 2.7-fold decrease in $G_{Na\_s}^{max}$ results in a 2.2-fold increase in the strength of the stimulating current that would still evoke a noxious reaction.

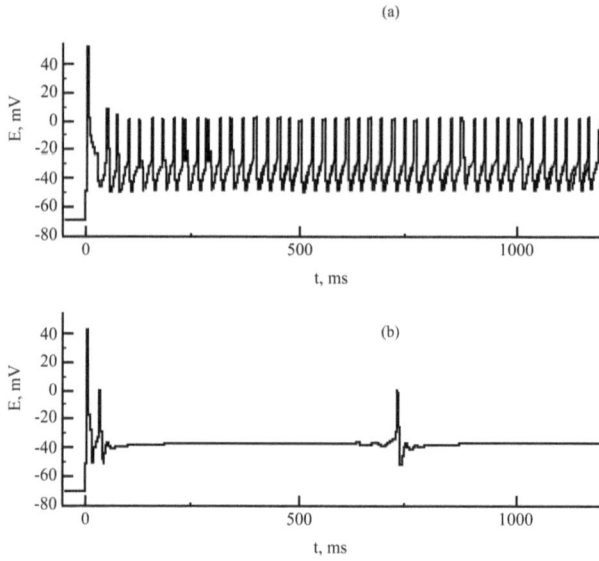

Fig. (1.11). Decrease of $Na_V1.8$ channel density results in a decrease of impulse frequency. $G_{Na\_s}^{max} = 200$ nS (a), $G_{Na\_s}^{max} = 75$ nS (b). $I_{stim} = 220$ pA.

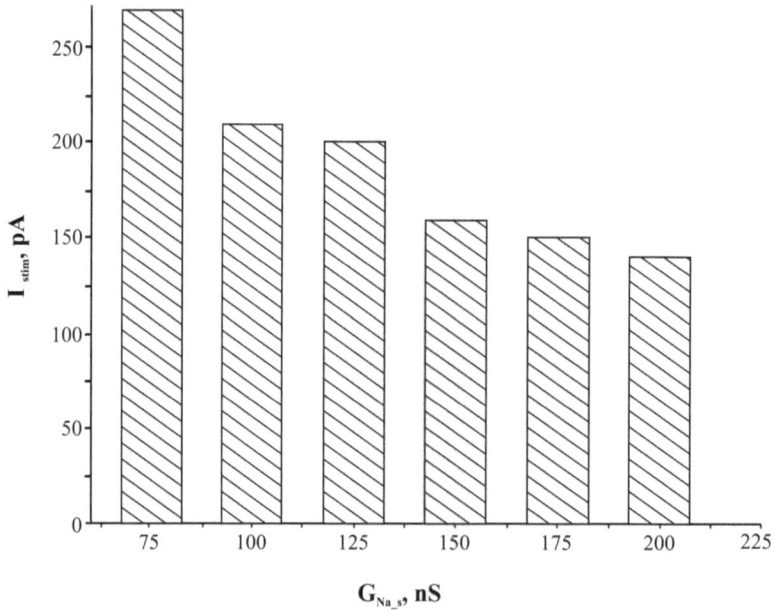

Fig. (1.12). $Na_V1.8$ channel density controls nociceptive signals.
Abscissa – maximum conductivity $G_{Na\_s}^{max}$;
Ordinate – the threshold values of stimulating currents that produce impulse firing with frequency exceeding 5 imp/s (the hypothetic frequency threshold of pain sensation).

As it may be concluded from the above, the following three mechanisms: 1) decrease of $Na_V1.8$ channel density; 2) positive shift of the steady-state activation voltage dependence; and 3) switching on slow inactivation process may diminish impulse firing frequency of nociceptive neuron given that the stimulus strength is constant. However, all these factors failed to specifically switch off the nociceptive response, since a sufficient increase of the stimulating current amplitude generates a high-frequency discharge characteristic of painful stimuli. On the contrary, only decrease of effective charge transfer specifically eliminates this high-frequency component responsible for pain sensation and only this modulating mechanism is independent on the stimulus strength.

PATCH-CLAMP RESULTS OF Z_{EFF} EVALUATION BY THE LIMITING LOGARITHMIC VOLTAGE SENSITIVITY METHOD

It is well known that opium as a painkiller has been used for around 6000 years. Its active ingredient, morphine, is still the most commonly prescribed medicine for relief of severe pain. This extremely potent analgesic was chosen as a primary object to test our experimental approach.

Morphine was shown to activate opioid receptors in the brain. There are four main types of opioid receptors that include μ, δ, κ receptors [194 - 196]. The fourth opioid receptor (NOP) also included in the opioid receptor family was termed the nociception orphanin FQ peptide receptor [197]. Opioid receptors are located widely throughout the central and peripheral nervous system. The highest densities of these receptors are located in the basal ganglia and thalamus, intermediate concentrations in the frontal and parietal cerebral cortex, and low concentrations in the cerebellum and occipital cortex [198]. Endogenous activation of opioid receptors is carried out by β-endorphin and endomorphin which have a particularly high affinity for these receptors. The target of morphine is well-studied μ-opioid receptor through which morphine exerts its analgesic and euphoric effects [199]. Morphine competes with endogenous neurotransmitters that activate receptors controlling the opioidergic system which has been implicated in a number of roles including analgesia, reward, sexual activity, learning and memory, and stress [200]. Opioids decrease pain sensitivity, reducing calcium-dependent release of neuromediators from presynaptic terminals, thus inhibiting the signal transmission between neurons [201 - 204].

All opioid receptors are coupled to G proteins which initiate a signaling cascade leading to decreased production of cyclic adenosine monophosphate (cAMP). The most important aspect of opioid receptor signal transduction relates to the ability of opioids to modulate calcium and potassium ion channels. Kir3 inwardly rectifying potassium channel modulation happens due to coupling of its gating device to G proteins [205, 206]. As a result, the activation curve of these channels is shifted to physiological range of E after opioid receptor has been switched on. That is why these channels decrease the rate of E growth during the interspike intervals, which leads to a strong inhibition of impulse firing.

When activated, opioid receptors also cause a reduction in calcium currents [207]. This effect can be explained by the decrease of their maximum conductivity. Opioid receptor-induced inhibition of calcium conductance is mediated by binding of the dissociated G protein subunit directly to the calcium channel. This binding event is also thought to reduce the voltage sensitivity of the channel [208 - 210], which might be accounted for by the positive shift of the activation voltage dependence of calcium channels along the E axis. These processes of potassium and calcium channels modulation should lead to a decrease in excitability of neuronal cells which is manifested in consequent pain relief.

In 1999 we have firstly shown that morphine is extremely effective for $Na_V1.8$ channels modulation [211]. It not only strongly inhibits the sodium current but also specifically decreases the effective charge of $Na_V1.8$ channel activation gating system. Application of the Almers' limiting slope method allowed to describe a radically novel mechanism of morphine action. Morphine was found to specifically activate a previously unknown opioid-like receptor as its molecular target with $K_d = 10$ nM, which is very close to the value obtained for morphine binding to classic μ-opioid receptor [212]. The role of signal transducer in this signaling pathway is played by Na^+,K^+-ATPase and not by G proteins. The signal triggered by morphine and strongly amplified by the transducer (Na^+,K^+-ATPase) finally reaches $Na_V1.8$ channel which serves as an effector unit. All these findings will be discussed in detail below. Our theoretical approach described above has demonstrated that the decrease of $Na_V1.8$ channel effective charge is a fine and delicate mechanism for physiologically adequate control of neuronal membrane firing. The gating characteristics of these channels are almost ideally suited for application of the Almers' method.

Prior to investigating morphine effects, it is necessary to determine experimental conditions of applicability of the limiting slope procedure which are based on the physical theory elegantly describing the reliability of the patch-clamp method [213]. The key parameter that controls the reliability of quantitative studies of changes in the effective charge of $Na_V1.8$ channel activation gating system is the series resistance R_s. The crucial fact of the theory is that R_s defines the dynamic and stationary error levels of the patch-clamp method [213] and also strongly influences the accuracy of Z_{eff} measurements. R_s was constantly monitored during all our patch-clamp experiments and maintained at values not exceeding 2-3 MΩ. If R_s is greater than 2-3 MΩ, the values of Z_{eff} are incorrect. The series resistance, in addition to limiting the accuracy of the method, also causes shifts in the characteristics being studied relative to the E axis. Stationary voltage shift (ΔE_{shift}) was evaluated by using the empirical formula $\Delta E_{shift} \cong I_{Na}^{max} \cdot R_s$. When I_{Na}^{max} was no greater than 1 nA, the error which would not exceed 2-3 mV was disregarded.

Our results provide evidence that application of morphine to the outer side of nociceptive neuron membrane strongly modulates the functioning of $Na_V1.8$ channels in a dose-dependent manner. Fig. (**1.13.a**) displays a family of $Na_V1.8$ currents after addition of morphine (0.1 μM) to extracellular solution. Application of morphine results in the decrease of both the amplitude values of sodium currents registered in all our experiments (see also [184]) and the effective charge value. Any marked changes in the kinetics of inactivation process were not detected. Morphine also strongly affects the activation gating system of $Na_V1.8$ channels inducing a positive shift of $G_{Na}^{max}(E)$ function along the E axis (Fig. **1.13.b**) that should lead to inhibition of impulse firing as discussed above. The most valuable effect obtained by the Almers' method is the detection of the decrease of Z_{eff} after morphine application as compared to the control data (Figs. **1.13.c** and **1.13.d**). The asymptotes introduced by Almers [58] in this case were plotted through the first 3 points at very negative E values. A more detailed analysis of the asymptote plotting yields almost the same Z_{eff} value (Fig. **1.14**), which demonstrates that Z_{eff} is independent on the sampling interval. The sampling step of 5 mV was chosen as an optimal one in all our further experiments, since it allowed to measure Z_{eff} with a sufficient accuracy and quickly enough while examined neuron was still viable.

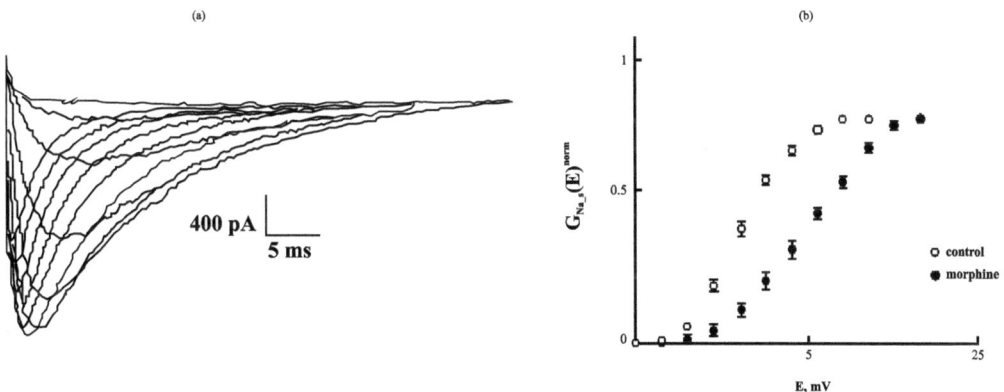

Fig. 1.13 contd.....

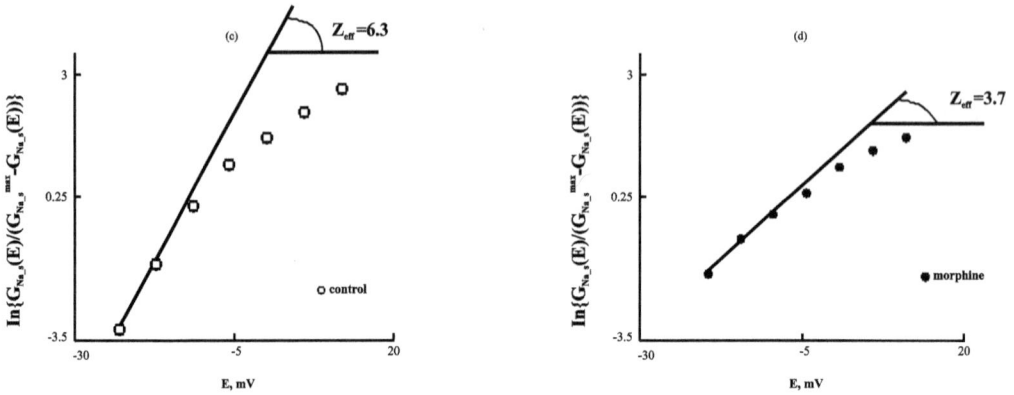

Fig. (1.13). Morphine effects on $Na_V1.8$ channels.

a – The family of $Na_V1.8$ currents after application of 0.1 μM morphine. Test voltage varied from -35 to 25 mV in square steps of 5 mV. Holding potential of 500 ms duration was -110 mV.

b – Voltage dependence of the normalized conductivity $G_{Na}^{norm}(E)$ of $Na_V1.8$ channels evaluated as $G_{Na}^{norm}(E) = G_{Na}(E)/G_{Na}^{max}(E)$), where $G_{Na}^{max}(E)$ is the maximal conductivity value. Data for control experiments (n = 23, white circles) and results obtained after application of 0.1 μM morphine (n = 27; black circles).

c – Evaluation of Z_{eff} in control experiments.

d – Evaluation of Z_{eff} after application of 0.1 μM morphine.

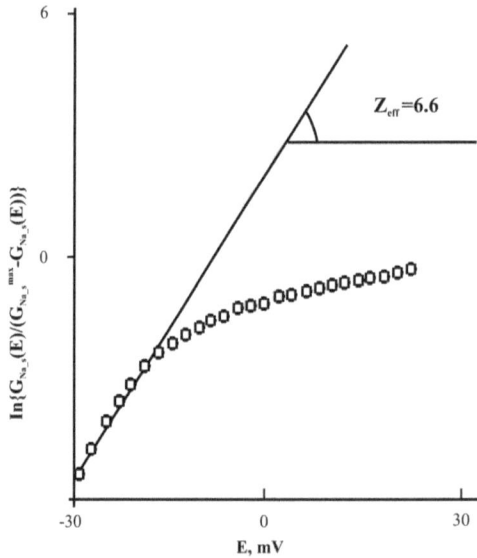

Fig. (1.14). Evaluation of Z_{eff} with reduced sampling interval.

The logarithmic voltage sensitivity function $L(E)$ makes it possible to determine Z_{eff} from the tangent of the slope of the asymptote plotted using the first 7 points when the voltage step was 2 mV (control experiment).

The most informative regarding Z_{eff} is the dose-effect function which is governed by molecular mechanism of modulation of $Na_V1.8$ channel gating machinery by morphine. This mechanism is considered in terms of the Hill equation that defines the physical basis of interaction between drug substances and their molecular targets.

Any agent applying for the role of a medicinal substance should have a specific mechanism of interaction with its molecular target. *A priori* it can be suggested that the more specific this mechanism is the less negative side effects should be produced by the agent, and this ligand-receptor interaction should improve the pathological state of human organism. The standard physiological approach to quantitatively describe the specificity of ligand-receptor interaction at the most detailed molecular level is to evaluate the equilibrium dissociation constant K_d (moles per liter, M) that might be defined by the Langmuir law:

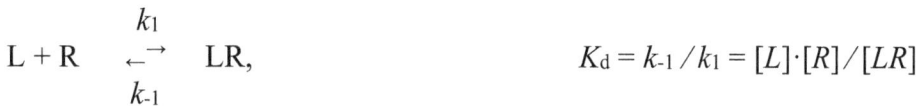

$$L + R \; \underset{k_{-1}}{\overset{k_1}{\rightleftharpoons}} \; LR, \qquad\qquad\qquad K_d = k_{-1}/k_1 = [L]\cdot[R]/[LR]$$

where $[L]$ is the ligand concentration, $[R]$ is its receptor concentration, $[LR]$ is the concentration of the ligand-receptor complex, brackets denote their equilibrium concentrations. k_1 and k_{-1} are the forward and backward rate constants, correspondingly. At equilibrium the fractional occupancy y of receptors is the saturation function of $[L]$ called the Langmuir adsorption isotherm:

$$y = [LR]/([LR] + [R]) = [L]/(K_d + [L]) = 1/(1 + (K_d/[L])).$$

The equations are similar to those applied in the Michaelis-Menten theory of enzyme kinetics. Half-maximal occupancy is attained when $[L]$, the ligand concentration, is numerically equal to the equilibrium dissociation constant K_d. It is easy to formulate the equation in order to calculate the fraction (r) of free receptors, $1 - y$:

$$r = 1 - y = K_d/(K_d + [L]) = 1/(1 + ([L]/K_d)).$$

This monotonic dose-response function makes it possible to obtain the K_d value of 5 nM for the inhibiting effect of saxitoxin binding. Response was equal to relative peak I_{Na} currents measured after different saxitoxin concentrations were added to the experimental bath [81].

In the case when the dose-response curve does not follow single-site kinetics as described by the Michaelis-Menten theory, the concentration dependence of ligand-receptor binding is better approximated by the Hill equation:

$$y = [L]^n / ((K_a)^n + [L]^n) = 1 / (1 + (K_a / [L])^n),$$

where y is the fraction of occupied ligand-binding sites on the receptor protein, n is the Hill coefficient, K_a is the ligand concentration producing half-maximal occupation of the binding sites. It should be noted that the Langmuir law is a particular case of the Hill equation with $n = 1$. The Hill equation describing the inhibiting dose-response function can be presented as

$$r = 1 - y = 1 / (1 + ([L]^n / K_d)).$$

Application of the Almers' method makes it possible to obtain Z_{eff} dependence on morphine concentration. Z_{eff} changed by roughly three elementary charges from $Z_{eff}^{max} = 5.9 \pm 0.5$ in control experiments to $Z_{eff}^{min} = 3.2 \pm 0.4$ after addition of morphine at a concentration of 10 μM. High morphine concentrations produced a saturation effect. It must be noted that in all our experiments Z_{eff} never reached zero. Should this happen, the ion channel would totally lose its voltage sensitivity. Our data allow to quantitatively describe the ligand-receptor binding process of morphine using the Hill function (Fig. **1.15**) and evaluate both the Hill coefficient ($n = 0.5$) and K_d (8 nM). We have also demonstrated that opioid receptors are not involved in effective charge transfer control. Intracellular exposure to an activator (GTP$_\gamma$S) and an inhibitor (GDP$_\beta$S) of G proteins and to pertussis toxin (a G$_i$ protein inhibitor) in conditions of extracellular application of 0.1 μM morphine did not influence the typical effect of morphine: Z_{eff} decreased by approximately two elementary charges. Thus, these agents had no effect on the process of signal transduction from the morphine target to Na$_V$1.8 channel activation gating device [211].

Our patch-clamp experiments show that a nonspecific opioid receptor blocker naloxone (NLX) applied at a concentration of 1 μM almost completely inhibits the effect of a decrease in Z_{eff} resulting from extracellular application of morphine (Fig. **1.16.a**). Exposure to naloxone alone led to a shift in the current-voltage characteristics by 5 ± 2 mV. Combined application of naloxone and morphine resulted in a twice as large shift in this characteristic ($\Delta E = 10 \pm 2$ mV) in the depolarizing direction (to the right). Fig. (**1.16.b**) indicates that application to extracellular solution of 100 μM ouabain, a specific blocker of Na$^+$,K$^+$-ATPase, nullified the effect of subsequently added morphine on Z_{eff}.

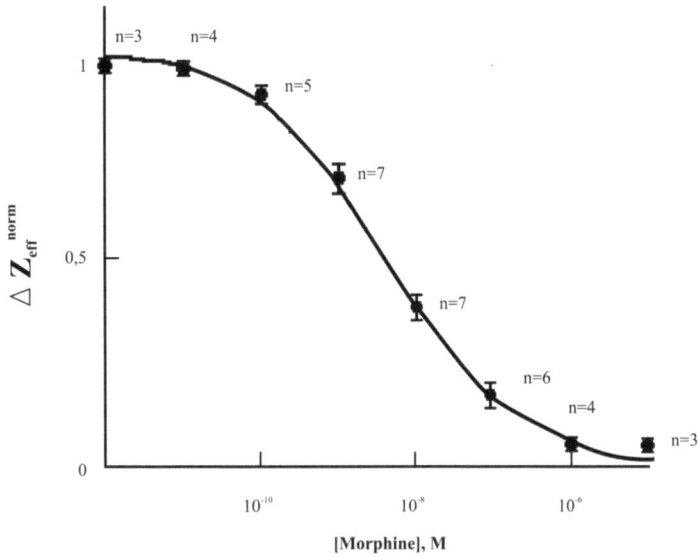

Fig. (1.15). Relationship between changes in the normalized effective charge (ΔZ_{eff}^{norm}) and extracellular morphine concentration.

$\Delta Z_{eff}^{norm} = (Z_{eff} - Z_{eff}^{min})/(Z_{eff}^{max} - Z_{eff}^{min})$. Mean values (black circles) and errors of the mean represent results obtained in several experiments, the numbers of which are shown beside each point on the plot. Solid line displays the results obtained by calculation using the Hill equation ($K_d = 8$ nM and $n = 0.5$).

(a)

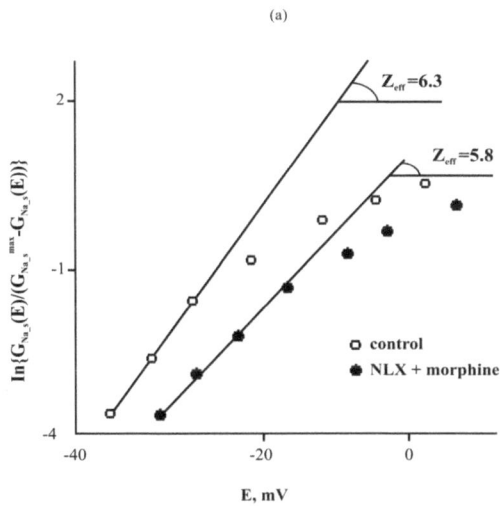

Fig. 1.16 contd.....

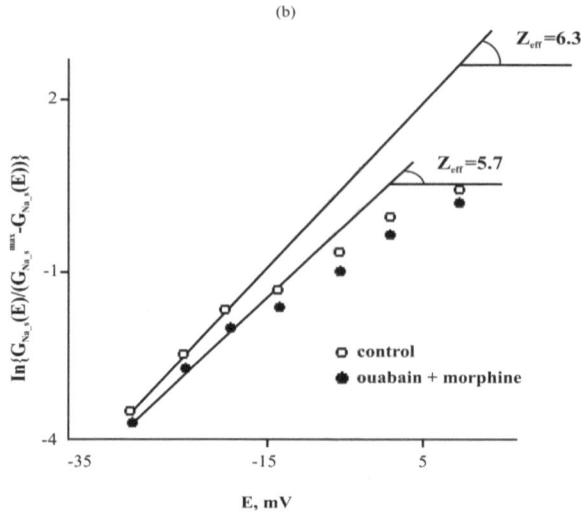

Fig. (1.16). Naloxone and ouabain nullify morphine-induced decrease in Z_{eff} of $Na_V1.8$ channel activation gating system.

a – White circles show values of the $L(E)$ function, data from the control experiments in the presence of naloxone only (1 μM); black circles show the $L(E)$ values obtained after combined application of morphine (20 μM) and naloxone (1 μM).

b – White circles show values of $L(E)$ function, data from the control experiments in the presence of ouabain (100 μM); black circles show $L(E)$ values obtained after combined application of morphine (20 μM) and ouabain (100 μM).

CONCLUSION

Summarizing the data obtained, one can conclude that morphine simultaneously activates three probable mechanisms leading to inhibition of impulse firing of nociceptive neuron membrane, all of which are under control of $Na_V1.8$ channels. This molecule might decrease the density of the channels; it shifts their activation gating function in the depolarizing direction and strongly diminishes the value of effective charge. It was shown above that each of these mechanisms might be a cause of pain relief due to inhibition of impulse firing. We thus hypothesize that modulation of $Na_V1.8$ channels by morphine is an additional mechanism of pain relief produced by this agent. Moreover, the fact that morphine activates all three background mechanisms that inhibit $Na_V1.8$ channels makes it an extremely potent analgesic. The agent has at least two molecular targets: a well-known μ-opioid receptor and as yet unidentified opioid-like receptor which we have described above using our physiological approach. It is tempting to predict that activation of the latter target does not evoke negative side effects as opposed to the case when the opioidergic system is activated. All the data on a novel mechanism of morphine action are summarized in Fig. (**1.17**). This molecule switches on the opioid-like receptor which, in turn, activates the transducer (Na^+,K^+-ATPase) connected as a series unit to $Na_v1.8$ channel. Our next task was to find an agent capable to specifically activate this new mechanism. It is very likely that such an agent should express potent analgesic properties, but it would not exhibit negative side effects of morphine.

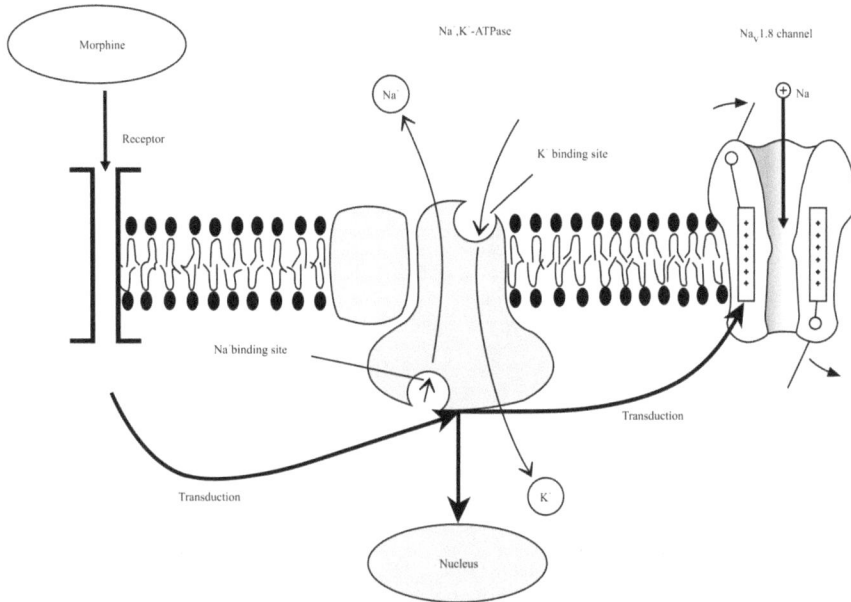

Fig. (1.17). Novel molecular mechanism of membrane signaling.
Three molecular structures participate in signal transduction after exposure to morphine: opioid-like receptor, Na^+,K^+-ATPase, and $Na_v1.8$ channel.

CONFLICT OF INTEREST

The authors confirm that they have no conflict of interest to declare for this publication.

ACKNOWLEDGEMENTS

Declared none.

REFERENCES

[1]   Adrian ED, Zotterman Y. The impulses produced by sensory nerve endings: Part 3. Impulses set up by touch and pressure. J Physiol 1926; 61(4): 465-83.
[http://dx.doi.org/10.1113/jphysiol.1926.sp002308] [PMID: 16993807]

[2]   Müller J. Handbuch der Physiologie des Menschen für Vorlesungen. 2 Bde. Coblenz: Hölscher 1844.

[3]   Sherrington C. The integrative action of the nervous system. 2nd ed., New Haven: Yale University Press 1947.

[4]   Mountcastle VB. The problem of sensing and the neural coding of sensory events. In: Quarton GC, Melnechuk T, Schmitt FO, Eds. The Neurosciences. New York: Rockefeller University Press 1967; pp. 393-408.

[5]   Mountcastle VB. Central nervous mechanisms in mechanoreceptive sensibility. In: Darian-Smith I, Ed. Handbook of physiology, Set 1, The nervous system, Vol III, Sensory processes, Pt 2. American Physiological Society 1984; pp. 789-878.

[6]   Somjen G. Sensory coding in the mammalian nervous system. New York: Appleton-Century-Crofts 1972.
[http://dx.doi.org/10.1007/978-1-4684-8190-7]

[7]   Goldscheider A. Das Schmerzproblem. Berlin: Springer 1920.
[http://dx.doi.org/10.1007/978-3-642-50863-9]

[8]   Pavlov IP. Conditioned reflexes: In: Anrep GV, Ed. an investigation of the physiological activity of the cerebral cortex GV Anrep (trans). London: Oxford University Press 1927.

[9]   Levin M. The pathogenesis of narcolepsy with a consideration of sleep-paralysis and localized sleep. J Neurol Psychopathol 1933; 14(53): 1-14.
[http://dx.doi.org/10.1136/jnnp.s1-14.53.1] [PMID: 21610755]

[10]   Loewenstein WR. Excitation and inactivation in a receptor membrane. Ann N Y Acad Sci 1961; 94(2): 510-34.
[http://dx.doi.org/10.1111/j.1749-6632.1961.tb35556.x] [PMID: 13763135]

[11]   Ilyinsky OB. Processes of excitation and inhibition in single mechanoreceptors (Pacinian corpuscles). Nature 1965; 208(5008): 351-3.
[http://dx.doi.org/10.1038/208351a0] [PMID: 5885446]

[12]   Ottoson D, Shepherd GM. Transducer properties and integrative mechanisms in the frog's muscle spindle. In: Loewenstein, Ed. Handbook of sensory physiology. Principles of receptor physiology. Berlin, Heidelberg: Springer-Verlag 1971; pp. 442-499.
[http://dx.doi.org/10.1007/978-3-642-65063-5_15]

[13]   Vallbo AB, Hagbarth KE. Activity from skin mechanoreceptors recorded percutaneously in awake human subjects. Exp Neurol 1968; 21(3): 270-89.
[http://dx.doi.org/10.1016/0014-4886(68)90041-1] [PMID: 5673644]

[14]   Vallbo AB, Hagbarth KE. Microelectrode recording from human peripheral nerves. In: Desmedt JE, Ed. New developments in electromyography and clinical neurophysiology . Basel, München, Paris, London, New York, Sydney: Karger 1973; pp. 67-84.
[http://dx.doi.org/10.1159/000394079]

[15]   Knibestöl M, Vallbo AB. Single unit analysis of mechanoreceptor activity from the human glabrous skin. Acta Physiol Scand 1970; 80(2): 178-95.
[http://dx.doi.org/10.1111/j.1748-1716.1970.tb04783.x] [PMID: 5475340]

[16]   Järvilehto T, Hämäläinen H, Laurinen P. Characteristics of single mechanoreceptive fibres innervating hairy skin of the human hand. Exp Brain Res 1976; 25(1): 45-61.
[http://dx.doi.org/10.1007/BF00237325] [PMID: 1269558]

[17]   Meyer RA, Campbell JN. Evidence for two distinct classes of unmyelinated nociceptive afferents in monkey. Brain Res 1981; 224(1): 149-52.
[http://dx.doi.org/10.1016/0006-8993(81)91124-0] [PMID: 7284829]

[18]   LaMotte RH, Thalhammer JG, Robinson CJ. Peripheral neural correlates of magnitude of cutaneous pain and hyperalgesia: a comparison of neural events in monkey with sensory judgments in human. J Neurophysiol 1983; 50(1): 1-26.
[PMID: 6875640]

[19]   Robinson CJ, Torebjörk HE, LaMotte RH. Psychophysical detection and pain ratings of incremental thermal stimuli: a comparison with nociceptor responses in humans. Brain Res 1983; 274(1): 87-106. [http://dx.doi.org/10.1016/0006-8993(83)90523-1] [PMID: 6616259]

[20]   Gybels J, Handwerker HO, Van Hees J. A comparison between the discharges of human nociceptive nerve fibres and the subjects ratings of his sensations. J Physiol 1979; 292(1): 193-206. [http://dx.doi.org/10.1113/jphysiol.1979.sp012846] [PMID: 490345]

[21]   LaMotte RH, Thalhammer JG, Torebjörk HE, Robinson CJ. Peripheral neural mechanisms of cutaneous hyperalgesia following mild injury by heat. J Neurosci 1982; 2(6): 765-81. [PMID: 7086482]

[22]   Torebjörk HE, LaMotte RH, Robinson CJ. Peripheral neural correlates of magnitude of cutaneous pain and hyperalgesia: simultaneous recordings in humans of sensory judgments of pain and evoked responses in nociceptors with C-fibers. J Neurophysiol 1984; 51(2): 325-39. [PMID: 6707724]

[23]   Meyer RA, Campbell JN. Myelinated nociceptive afferents account for the hyperalgesia that follows a burn to the hand. Science 1981; 213(4515): 1527-9. [http://dx.doi.org/10.1126/science.7280675] [PMID: 7280675]

[24]   LaMotte RH, Lundberg LE, Torebjörk HE. Pain, hyperalgesia and activity in nociceptive C units in humans after intradermal injection of capsaicin. J Physiol 1992; 448(1): 749-64. [http://dx.doi.org/10.1113/jphysiol.1992.sp019068] [PMID: 1593488]

[25]   Willis WD, Sluka KA, Rees H, Westlund KN. Cooperative mechanisms of neurotransmitter action in central nervous sensitization. Prog Brain Res 1996; 110(110): 151-66. [http://dx.doi.org/10.1016/S0079-6123(08)62572-8] [PMID: 9000723]

[26]   Besson JM, Chaouch A. Peripheral and spinal mechanisms of nociception. Physiol Rev 1987; 67(1): 67-186. [PMID: 3543978]

[27]   Treede R-D, Meyer RA, Raja SN, Campbell JN. Peripheral and central mechanisms of cutaneous hyperalgesia. Prog Neurobiol 1992; 38(4): 397-421. [http://dx.doi.org/10.1016/0301-0082(92)90027-C] [PMID: 1574584]

[28]   Koltzenburg M, Handwerker HO. Differential ability of human cutaneous nociceptors to signal mechanical pain and to produce vasodilatation. J Neurosci 1994; 14(3 Pt 2): 1756-65. [PMID: 8126568]

[29]   Konietzny F, Hensel H. The dynamic response of warm units in human skin nerves. Pflügers Arch 1977; 370(1): 111-4. [http://dx.doi.org/10.1007/BF00707956] [PMID: 561379]

[30]   Molinari HH, Kenshalo DR. Effect of cooling rate on the dynamic response of cat cold units. Exp Neurol 1977; 55(3 Pt. 1): 546-55. [http://dx.doi.org/10.1016/0014-4886(77)90283-7] [PMID: 300684]

[31]   Van Hees J, Gybels J. C nociceptor activity in human nerve during painful and non painful skin stimulation. J Neurol Neurosurg Psychiatry 1981; 44(7): 600-7. [http://dx.doi.org/10.1136/jnnp.44.7.600] [PMID: 7288447]

[32]   Adriaensen H, Gybels J, Handwerker HO, Van Hees J. Nociceptor discharges and sensations due to prolonged noxious mechanical stimulation - a paradox. Hum Neurobiol 1984; 3(1): 53-8. [PMID: 6330012]

[33]   Adriaensen H, Gybels J, Handwerker HO, Van Hees J. Latencies of chemically evoked discharges in human cutaneous nociceptors and of the concurrent subjective sensations. Neurosci Lett 1980; 20(1): 55-9. [http://dx.doi.org/10.1016/0304-3940(80)90233-5] [PMID: 7052549]

[34]   Lynn B, Cotsell B. The delay in onset of vasodilator flare in human skin at increasing distances from a localized noxious stimulus. Microvasc Res 1991; 41(2): 197-202.
[http://dx.doi.org/10.1016/0026-2862(91)90021-3] [PMID: 1828855]

[35]   Magerl W, Szolcsányi J, Westerman RA, Handwerker HO. Laser Doppler measurements of skin vasodilation elicited by percutaneous electrical stimulation of nociceptors in humans. Neurosci Lett 1987; 82(3): 349-54.
[http://dx.doi.org/10.1016/0304-3940(87)90281-3] [PMID: 2962019]

[36]   Szolcsányi J. Antidromic vasodilatation and neurogenic inflammation. Agents Actions 1988; 23(1-2): 4-11.
[http://dx.doi.org/10.1007/BF01967170] [PMID: 2965495]

[37]   Magerl W, Westerman RA, Möhner B, Handwerker HO. Properties of transdermal histamine iontophoresis: differential effects of season, gender, and body region. J Invest Dermatol 1990; 94(3): 347-52.
[http://dx.doi.org/10.1111/1523-1747.ep12874474] [PMID: 2307854]

[38]   Lundberg LE, Jørum E, Holm E, Torebjörk HE. Intra-neural electrical stimulation of cutaneous nociceptive fibres in humans: effects of different pulse patterns on magnitude of pain. Acta Physiol Scand 1992; 146(1): 41-8.
[http://dx.doi.org/10.1111/j.1748-1716.1992.tb09391.x] [PMID: 1442126]

[39]   Wall PD, Woolf CJ. Muscle but not cutaneous C-afferent input produces prolonged increases in the excitability of the flexion reflex in the rat. J Physiol 1984; 356(1): 443-58.
[http://dx.doi.org/10.1113/jphysiol.1984.sp015475] [PMID: 6520794]

[40]   Willis WD, Coggeshall RE. Sensory function of the spinal cord. 2nd ed., New York: Plenum 1991.
[http://dx.doi.org/10.1007/978-1-4899-0597-0]

[41]   Woolf CJ. Excitability changes in central neurons following peripheral damage: role of central sensitization in the pathogenesis of pain. In: Willis WD, Ed. Hyperalgesia and allodynia. New York: Raven 1992; pp. 221-43.

[42]   Dubner R, Ruda MA. Activity-dependent neuronal plasticity following tissue injury and inflammation. Trends Neurosci 1992; 15(3): 96-103.
[http://dx.doi.org/10.1016/0166-2236(92)90019-5] [PMID: 1373925]

[43]   Wall PD, Melzack R, Eds. Textbook of Pain. 2nd ed., Edinburgh: Churchhill Livingstone 1989.

[44]   Wall PD, Devor M. Sensory afferent impulses originate from dorsal root ganglia as well as from the periphery in normal and nerve injured rats. Pain 1983; 17(4): 321-39.
[http://dx.doi.org/10.1016/0304-3959(83)90164-1] [PMID: 6664680]

[45]   Devor M. Unexplained peculiarities of the dorsal root ganglion. Pain 1999; (Suppl. 6)S27-35.
[http://dx.doi.org/10.1016/S0304-3959(99)00135-9] [PMID: 10491970]

[46]   Kirk EJ. Impulses in dorsal spinal nerve rootlets in cats and rabbits arising from dorsal root ganglia isolated from the periphery. J Comp Neurol 1974; 155(2): 165-75.
[http://dx.doi.org/10.1002/cne.901550203] [PMID: 4827008]

[47]   Burchiel KJ. Effects of electrical and mechanical stimulation on two foci of spontaneous activity which develop in primary afferent neurons after peripheral axotomy. Pain 1984; 18(3): 249-65.
[http://dx.doi.org/10.1016/0304-3959(84)90820-0] [PMID: 6728494]

[48]   Nordin M, Nyström B, Wallin U, Hagbarth K-E. Ectopic sensory discharges and paresthesiae in patients with disorders of peripheral nerves, dorsal roots and dorsal columns. Pain 1984; 20(3): 231-45.
[http://dx.doi.org/10.1016/0304-3959(84)90013-7] [PMID: 6096790]

[49]   Kuslich SD, Ulstrom CL, Michael CJ. The tissue origin of low back pain and sciatica: a report of pain response to tissue stimulation during operations on the lumbar spine using local anesthesia. Orthop

Clin North Am 1991; 22(2): 181-7.
[PMID: 1826546]

[50]   Kajander KC, Wakisaka S, Bennett GJ. Spontaneous discharge originates in the dorsal root ganglion at the onset of a painful peripheral neuropathy in the rat. Neurosci Lett 1992; 138(2): 225-8.
        [http://dx.doi.org/10.1016/0304-3940(92)90920-3] [PMID: 1319012]

[51]   Sheen K, Chung JM. Signs of neuropathic pain depend on signals from injured nerve fibers in a rat model. Brain Res 1993; 610(1): 62-8.
        [http://dx.doi.org/10.1016/0006-8993(93)91217-G] [PMID: 8518931]

[52]   Devor M, Seltzer Z. Pathophysiology of damaged nerves in relation to chronic pain. In: Wall PD, Melzack R, Eds. Textbook of pain 4. London: Churchill Livingstone 1999; pp. 129-64.

[53]   Liu C-N, Michaelis M, Amir R, Devor M. Spinal nerve injury enhances subthreshold membrane potential oscillations in DRG neurons: relation to neuropathic pain. J Neurophysiol 2000; 84(1): 205-15.
        [PMID: 10899197]

[54]   Liu C-N, Devor M, Waxman SG, Kocsis JD. Subthreshold oscillations induced by spinal nerve injury in dissociated muscle and cutaneous afferents of mouse DRG. J Neurophysiol 2002; 87(4): 2009-17.
        [http://dx.doi.org/10.1152/jn.00705.2001] [PMID: 11929919]

[55]   Boucher TJ, McMahon SB. Neurotrophic factors and neuropathic pain. Curr Opin Pharmacol 2001; 1(1): 66-72.
        [http://dx.doi.org/10.1016/S1471-4892(01)00010-8] [PMID: 11712538]

[56]   Sukhotinsky I, Ben-Dor E, Raber P, Devor M. Key role of the dorsal root ganglion in neuropathic tactile hypersensibility. Eur J Pain 2004; 8(2): 135-43.
        [http://dx.doi.org/10.1016/S1090-3801(03)00086-7] [PMID: 14987623]

[57]   Amir R, Michaelis M, Devor M. Burst discharge in primary sensory neurons: triggered by subthreshold oscillations, maintained by depolarizing afterpotentials. J Neurosci 2002; 22(3): 1187-98.
        [PMID: 11826148]

[58]   Almers W. Gating currents and charge movements in excitable membranes. Rev Physiol Biochem Pharmacol 1978; 82: 96-190.
        [http://dx.doi.org/10.1007/BFb0030498] [PMID: 356157]

[59]   Hodgkin AL, Huxley AF. Currents carried by sodium and potassium ions through the membrane of the giant axon of *Loligo*. J Physiol 1952; 116(4): 449-72.
        [http://dx.doi.org/10.1113/jphysiol.1952.sp004717] [PMID: 14946713]

[60]   Hodgkin AL, Huxley AF. The components of membrane conductance in the giant axon of *Loligo*. J Physiol 1952; 116(4): 473-96.
        [http://dx.doi.org/10.1113/jphysiol.1952.sp004718] [PMID: 14946714]

[61]   Hodgkin AL, Huxley AF. The dual effect of membrane potential on sodium conductance in the giant axon of *Loligo*. J Physiol 1952; 116(4): 497-506.
        [http://dx.doi.org/10.1113/jphysiol.1952.sp004719] [PMID: 14946715]

[62]   Hodgkin AL, Huxley AF. A quantitative description of membrane current and its application to conduction and excitation in nerve. J Physiol 1952; 117(4): 500-44.
        [http://dx.doi.org/10.1113/jphysiol.1952.sp004764] [PMID: 12991237]

[63]   Hodgkin AL, Keynes RD. The potassium permeability of a giant nerve fibre. J Physiol 1955; 128(1): 61-88.
        [http://dx.doi.org/10.1113/jphysiol.1955.sp005291] [PMID: 14368575]

[64]   Noda M, Shimizu S, Tanabe T, *et al.* Primary structure of *Electrophorus electricus* sodium channel deduced from cDNA sequence. Nature 1984; 312(5990): 121-7.
        [http://dx.doi.org/10.1038/312121a0] [PMID: 6209577]

[65] Krylov BV, Makovsky VS. Spike frequency adaptation in amphibian sensory fibres is probably due to slow K channels. Nature 1978; 275(5680): 549-51.
[http://dx.doi.org/10.1038/275549a0] [PMID: 692733]

[66] Chiu SY, Ritchie JM, Rogart RB, Stagg D. A quantitative description of membrane currents in rabbit myelinated nerve. J Physiol 1979; 292(1): 149-66.
[http://dx.doi.org/10.1113/jphysiol.1979.sp012843] [PMID: 314974]

[67] Chiu SY, Ritchie JM. Evidence for the presence of potassium channels in the paranodal region of acutely demyelinated mammalian single nerve fibres. J Physiol 1981; 313(1): 415-37.
[http://dx.doi.org/10.1113/jphysiol.1981.sp013674] [PMID: 6268773]

[68] Akoev GN, Makovsky VS, Volpe NO. Effects of tetraethylammonium on mechano- and electrosensitive channels of Pacinian corpuscle. Neurosci Lett 1980; 19(1): 61-6.
[http://dx.doi.org/10.1016/0304-3940(80)90256-6] [PMID: 6302593]

[69] Akoev GN, Alekseev NP, Krylov BV. Mechanoreceptors: Their functional organization. London, Heidelberg, N-Y: Springer-Verlag 1988.
[http://dx.doi.org/10.1007/978-3-642-72935-5]

[70] Widmark J, Sundström G, Ocampo Daza D, Larhammar D. Differential evolution of voltage-gated sodium channels in tetrapods and teleost fishes. Mol Biol Evol 2011; 28(1): 859-71.
[http://dx.doi.org/10.1093/molbev/msq257] [PMID: 20924084]

[71] Zakon HH, Jost MC, Lu Y. Expansion of voltage-dependent Na+ channel gene family in early tetrapods coincided with the emergence of terrestriality and increased brain complexity. Mol Biol Evol 2011; 28(4): 1415-24.
[http://dx.doi.org/10.1093/molbev/msq325] [PMID: 21148285]

[72] Catterall WA, Goldin AL, Waxman SG. International Union of Pharmacology. XLVII. Nomenclature and structure-function relationships of voltage-gated sodium channels. Pharmacol Rev 2005; 57(4): 397-409.
[http://dx.doi.org/10.1124/pr.57.4.4] [PMID: 16382098]

[73] Yu FH, Catterall WA. Overview of the voltage-gated sodium channel family. Genome Biol 2003; 4(3): 207.
[http://dx.doi.org/10.1186/gb-2003-4-3-207] [PMID: 12620097]

[74] Yu FH, Yarov-Yarovoy V, Gutman GA, Catterall WA. Overview of molecular relationships in the voltage-gated ion channel superfamily. Pharmacol Rev 2005; 57(4): 387-95.
[http://dx.doi.org/10.1124/pr.57.4.13] [PMID: 16382097]

[75] Catterall WA. From ionic currents to molecular mechanisms: the structure and function of voltage-gated sodium channels. Neuron 2000; 26(1): 13-25.
[http://dx.doi.org/10.1016/S0896-6273(00)81133-2] [PMID: 10798388]

[76] Catterall WA. Structure and function of voltage-gated sodium channels at atomic resolution. Exp Physiol 2014; 99(1): 35-51.
[http://dx.doi.org/10.1113/expphysiol.2013.071969] [PMID: 24097157]

[77] Brackenbury WJ, Calhoun JD, Chen C, *et al.* Functional reciprocity between Na$^+$ channel Na$_v$1.6 and β1 subunits in the coordinated regulation of excitability and neurite outgrowth. Proc Natl Acad Sci USA 2010; 107(5): 2283-8.
[http://dx.doi.org/10.1073/pnas.0909434107] [PMID: 20133873]

[78] Brackenbury WJ, Isom LL. Na$^+$ channel β subunits: overachievers of the ion channel family. Front Pharmacol 2011; 2: 53.
[http://dx.doi.org/10.3389/fphar.2011.00053] [PMID: 22007171]

[79] Mullins LJ. The macromolecular properties of excitable membranes. Ann N Y Acad Sci 1961; 94(2): 390-404.
[http://dx.doi.org/10.1111/j.1749-6632.1961.tb35553.x] [PMID: 13726754]

[80]   Eisenman G, Sandblom JP, Walker JL Jr. Membrane structure and ion permeation. Study of ion exchange membrane structure and function is relevant to analysis of biological ion permeation. Science 1967; 155(3765): 965-74.
[http://dx.doi.org/10.1126/science.155.3765.965] [PMID: 5334938]

[81]   Hille B. Pharmacological modifications of the sodium channels of frog nerve. J Gen Physiol 1968; 51(2): 199-219.
[http://dx.doi.org/10.1085/jgp.51.2.199] [PMID: 5641635]

[82]   Hille B. The permeability of the sodium channel to organic cations in myelinated nerve. J Gen Physiol 1971; 58(6): 599-619.
[http://dx.doi.org/10.1085/jgp.58.6.599] [PMID: 5315827]

[83]   Hille B. Potassium channels in myelinated nerve. Selective permeability to small cations. J Gen Physiol 1973; 61(6): 669-86.
[http://dx.doi.org/10.1085/jgp.61.6.669] [PMID: 4541077]

[84]   Guidoni L, Torre V, Carloni P. Potassium and sodium binding to the outer mouth of the K+ channel. Biochemistry 1999; 38(27): 8599-604.
[http://dx.doi.org/10.1021/bi990540c] [PMID: 10393534]

[85]   Boda D, Nonner W, Valiskó M, Henderson D, Eisenberg B, Gillespie D. Steric selectivity in Na channels arising from protein polarization and mobile side chains. Biophys J 2007; 93(6): 1960-80.
[http://dx.doi.org/10.1529/biophysj.107.105478] [PMID: 17526571]

[86]   Noskov SY, Bernèche S, Roux B. Control of ion selectivity in potassium channels by electrostatic and dynamic properties of carbonyl ligands. Nature 2004; 431(7010): 830-4.
[http://dx.doi.org/10.1038/nature02943] [PMID: 15483608]

[87]   Varma S, Rempe SB. Tuning ion coordination architectures to enable selective partitioning. Biophys J 2007; 93(4): 1093-9.
[http://dx.doi.org/10.1529/biophysj.107.107482] [PMID: 17513348]

[88]   Blumenthal KM. Ion channels as targets for toxins. In: Sperelakis N, Ed. Cell physiology sourcebook: essentials of membrane biophysics. 4th ed. Amsterdam, Boston: Elsevier/Academic press 2012; pp. 509-24.
[http://dx.doi.org/10.1016/B978-0-12-387738-3.00028-7]

[89]   Hille B, Ed. Ionic channels of excitable membranes. 3rd ed., Sunderland: Sinauer Associates 2001.

[90]   Wang S-Y, Mitchell J, Tikhonov DB, Zhorov BS, Wang GK. How batrachotoxin modifies the sodium channel permeation pathway: computer modeling and site-directed mutagenesis. Mol Pharmacol 2006; 69(3): 788-95.
[PMID: 16354762]

[91]   Khodorov BI. Chemicals as tools to study nerve fibre sodium channels. Effects of batrachotoxin and some local anesthetics. In: Tosteson D, Ovchinnikov Yu, Lattorre R, Eds. Membrane transport processes. New York: Raven Press 1978; pp. 153-74.

[92]   Catterall WA. Structure and function of voltage-gated ion channels. Annu Rev Biochem 1995; 64(1): 493-531.
[http://dx.doi.org/10.1146/annurev.bi.64.070195.002425] [PMID: 7574491]

[93]   Moczydlowski E, Olivera BM, Gray WR, Strichartz GR. Discrimination of muscle and neuronal Na-channel subtypes by binding competition between [3H]saxitoxin and μ-conotoxins. Proc Natl Acad Sci USA 1986; 83(14): 5321-5.
[http://dx.doi.org/10.1073/pnas.83.14.5321] [PMID: 2425365]

[94]   Ritchie JM, Rogart RB. The binding of saxitoxin and tetrodotoxin to excitable tissue. Rev Physiol Biochem Pharmacol 1977; 79: 1-50.
[http://dx.doi.org/10.1007/BFb0037088] [PMID: 335473]

[95] Kostyuk PG, Veselovsky NS, Tsyndrenko AY. Ionic currents in the somatic membrane of rat dorsal root ganglion neurons - I. Sodium currents. Neuroscience 1981; 6(12): 2423-30.
[http://dx.doi.org/10.1016/0306-4522(81)90088-9] [PMID: 6275294]

[96] Gold MS, Reichling DB, Shuster MJ, Levine JD. Hyperalgesic agents increase a tetrodotoxin-resistant Na+ current in nociceptors. Proc Natl Acad Sci USA 1996; 93(3): 1108-12.
[http://dx.doi.org/10.1073/pnas.93.3.1108] [PMID: 8577723]

[97] Borovikova L, Borovikov D, Ermishkin V, Revenko S. The resistance of cutaneous feline C-fiber mechano-heat-sensitive unit termination to tetrodotoxin and its possible relation to tetrodotoxin-resistant sodium channels. Prim Sens Neuron 1997; 2(1): 65-75.
[http://dx.doi.org/10.1163/092996397750131883]

[98] Catterall WA. Structure and function of voltage-sensitive ion channels. Science 1988; 242(4875): 50-61.
[http://dx.doi.org/10.1126/science.2459775] [PMID: 2459775]

[99] Terlau H, Heinemann SH, Stühmer W, *et al.* Mapping the site of block by tetrodotoxin and saxitoxin of sodium channel II. FEBS Lett 1991; 293(1-2): 93-6.
[http://dx.doi.org/10.1016/0014-5793(91)81159-6] [PMID: 1660007]

[100] Satin J, Limberis JT, Kyle JW, Rogart RB, Fozzard HA. The saxitoxin/tetrodotoxin binding site on cloned rat brain IIa Na channels is in the transmembrane electric field. Biophys J 1994; 67(3): 1007-14.
[http://dx.doi.org/10.1016/S0006-3495(94)80566-1] [PMID: 7811911]

[101] Noda M, Suzuki H, Numa S, Stühmer W. A single point mutation confers tetrodotoxin and saxitoxin insensitivity on the sodium channel II. FEBS Lett 1989; 259(1): 213-6.
[http://dx.doi.org/10.1016/0014-5793(89)81531-5] [PMID: 2557243]

[102] Kontis KJ, Goldin AL. Site-directed mutagenesis of the putative pore region of the rat IIA sodium channel. Mol Pharmacol 1993; 43(4): 635-44.
[PMID: 8386312]

[103] Stephan MM, Potts JF, Agnew WS. The μI skeletal muscle sodium channel: mutation E403Q eliminates sensitivity to tetrodotoxin but not to μ-conotoxins GIIIA and GIIIB. J Membr Biol 1994; 137(1): 1-8.
[http://dx.doi.org/10.1007/BF00234993] [PMID: 7911843]

[104] Khodorov BI, Shishkova LD, Peganov EM, Revenko SV. Inhibition of sodium currents in frog Ranvier node treated with local anesthetics. Role of slow sodium inactivation. Biochim Biophys Acta 1976; 433(2): 409-35.
[http://dx.doi.org/10.1016/0005-2736(76)90105-X] [PMID: 1260035]

[105] Khodorov B. Some aspects of the pharmacology of sodium channels in nerve membrane. Process of inactivation. Biochem Pharmacol 1979; 28(9): 1451-9.
[http://dx.doi.org/10.1016/0006-2952(79)90457-X] [PMID: 383084]

[106] Khodorov BI. Sodium inactivation and drug-induced immobilization of the gating charge in nerve membrane. Prog Biophys Mol Biol 1981; 37(2): 49-89.
[http://dx.doi.org/10.1016/0079-6107(82)90020-7] [PMID: 6264546]

[107] Hille B. Local anesthetics: hydrophilic and hydrophobic pathways for the drug-receptor reaction. J Gen Physiol 1977; 69(4): 497-515.
[http://dx.doi.org/10.1085/jgp.69.4.497] [PMID: 300786]

[108] Mozhayeva GN, Naumov AP. Effect of surface charge on the steady-state potassium conductance of nodal membrane. Nature 1970; 228(5267): 164-5.
[http://dx.doi.org/10.1038/228164a0] [PMID: 5460015]

[109] Mozhayeva GN, Naumov AP. The permeability of sodium channels to hydrogen ions in nerve fibres. Pflügers Arch 1983; 396(2): 163-73.

[http://dx.doi.org/10.1007/BF00615521] [PMID: 6300755]

[110] Mozhaeva GN, Naumov AP, Neguliaev IuA. [The influence of aconitine on several properties of the sodium channels of the membranes of nodes of Ranvier]. Neirofiziologiia 1976; 8(2): 152-60.
[PMID: 1272460]

[111] Mozhaeva GN, Naumov AP, Nosyreva ED, Grishin EV. Potential-dependent interaction of toxin from venom of the scorpion *Buthus eupeus* with sodium channels in myelinated fiber. Biochim Biophys Acta 1980; 597(3): 587-602.
[http://dx.doi.org/10.1016/0005-2736(80)90230-8] [PMID: 6246941]

[112] Armstrong CM, Bezanilla F, Rojas E. Destruction of sodium conductance inactivation in squid axons perfused with pronase. J Gen Physiol 1973; 62(4): 375-91.
[http://dx.doi.org/10.1085/jgp.62.4.375] [PMID: 4755846]

[113] Stühmer W, Conti F, Suzuki H, *et al.* Structural parts involved in activation and inactivation of the sodium channel. Nature 1989; 339(6226): 597-603.
[http://dx.doi.org/10.1038/339597a0] [PMID: 2543931]

[114] Vassilev P, Scheuer T, Catterall WA. Inhibition of inactivation of single sodium channels by a site-directed antibody. Proc Natl Acad Sci USA 1989; 86(20): 8147-51.
[http://dx.doi.org/10.1073/pnas.86.20.8147] [PMID: 2554301]

[115] Joseph D, Petsko GA, Karplus M. Anatomy of a conformational change: hinged lid motion of the triosephosphate isomerase loop. Science 1990; 249(4975): 1425-8.
[http://dx.doi.org/10.1126/science.2402636] [PMID: 2402636]

[116] Rohl CA, Boeckman FA, Baker C, Scheuer T, Catterall WA, Klevit RE. Solution structure of the sodium channel inactivation gate. Biochemistry 1999; 38(3): 855-61.
[http://dx.doi.org/10.1021/bi9823380] [PMID: 9893979]

[117] Kellenberger S, West JW, Scheuer T, Catterall WA. Molecular analysis of the putative inactivation particle in the inactivation gate of brain type IIA Na^+ channels. J Gen Physiol 1997; 109(5): 589-605.
[http://dx.doi.org/10.1085/jgp.109.5.589] [PMID: 9154906]

[118] West JW, Patton DE, Scheuer T, Wang Y, Goldin AL, Catterall WA. A cluster of hydrophobic amino acid residues required for fast $Na^{(+)}$-channel inactivation. Proc Natl Acad Sci USA 1992; 89(22): 10910-4.
[http://dx.doi.org/10.1073/pnas.89.22.10910] [PMID: 1332060]

[119] Sarhan MF, Tung CC, Van Petegem F, Ahern CA. Crystallographic basis for calcium regulation of sodium channels. Proc Natl Acad Sci USA 2012; 109(9): 3558-63.
[http://dx.doi.org/10.1073/pnas.1114748109] [PMID: 22331908]

[120] Wang SY, Bonner K, Russell C, Wang GK. Tryptophan scanning of D1S6 and D4S6 C-termini in voltage-gated sodium channels. Biophys J 2003; 85(2): 911-20.
[http://dx.doi.org/10.1016/S0006-3495(03)74530-5] [PMID: 12885638]

[121] Ahern CA, Payandeh J, Bosmans F, Chanda B. The hitchhikers guide to the voltage-gated sodium channel galaxy. J Gen Physiol 2016; 147(1): 1-24.
[http://dx.doi.org/10.1085/jgp.201511492] [PMID: 26712848]

[122] Guy HR, Seetharamulu P. Molecular model of the action potential sodium channel. Proc Natl Acad Sci USA 1986; 83(2): 508-12.
[http://dx.doi.org/10.1073/pnas.83.2.508] [PMID: 2417247]

[123] Catterall WA. Voltage-dependent gating of sodium channels: correlating structure and function. Trends Neurosci 1986; 9(1): 7-10.
[http://dx.doi.org/10.1016/0166-2236(86)90004-4]

[124] Yang N, George AL Jr, Horn R. Molecular basis of charge movement in voltage-gated sodium channels. Neuron 1996; 16(1): 113-22.
[http://dx.doi.org/10.1016/S0896-6273(00)80028-8] [PMID: 8562074]

[125] Yarov-Yarovoy V, Baker D, Catterall WA. Voltage sensor conformations in the open and closed states in ROSETTA structural models of K($^+$) channels. Proc Natl Acad Sci USA 2006; 103(19): 7292-7.
[http://dx.doi.org/10.1073/pnas.0602350103] [PMID: 16648251]

[126] Yarov-Yarovoy V, DeCaen PG, Westenbroek RE, *et al.* Structural basis for gating charge movement in the voltage sensor of a sodium channel. Proc Natl Acad Sci USA 2012; 109(2): E93-E102.
[http://dx.doi.org/10.1073/pnas.1118434109] [PMID: 22160714]

[127] Starace DM, Bezanilla F. A proton pore in a potassium channel voltage sensor reveals a focused electric field. Nature 2004; 427(6974): 548-53.
[http://dx.doi.org/10.1038/nature02270] [PMID: 14765197]

[128] Chanda B, Asamoah OK, Blunck R, Roux B, Bezanilla F. Gating charge displacement in voltage-gated ion channels involves limited transmembrane movement. Nature 2005; 436(7052): 852-6.
[http://dx.doi.org/10.1038/nature03888] [PMID: 16094369]

[129] DeCaen PG, Yarov-Yarovoy V, Scheuer T, Catterall WA. Gating charge interactions with the S1 segment during activation of a Na$^+$ channel voltage sensor. Proc Natl Acad Sci USA 2011; 108(46): 18825-30.
[http://dx.doi.org/10.1073/pnas.1116449108] [PMID: 22042870]

[130] DeCaen PG, Yarov-Yarovoy V, Sharp EM, Scheuer T, Catterall WA. Sequential formation of ion pairs during activation of a sodium channel voltage sensor. Proc Natl Acad Sci USA 2009; 106(52): 22498-503.
[http://dx.doi.org/10.1073/pnas.0912307106] [PMID: 20007787]

[131] Chandler WK, Meves H. Sodium and potassium currents in squid axons perfused with fluoride solutions. J Physiol 1970; 211(3): 623-52.
[http://dx.doi.org/10.1113/jphysiol.1970.sp009297] [PMID: 5501055]

[132] Braun AP. Cooperative interactions between subunits regulate gating in holo-proton conductive channels. Channels (Austin) 2010; 4(2): 73-4.
[http://dx.doi.org/10.4161/chan.4.2.12111] [PMID: 21228637]

[133] Liin SI, Barro-Soria R, Larsson HP. The KCNQ1 channel - remarkable flexibility in gating allows for functional versatility. J Physiol 2015; 593(12): 2605-15.
[http://dx.doi.org/10.1113/jphysiol.2014.287607] [PMID: 25653179]

[134] Chevrier P, Vijayaragavan K, Chahine M. Differential modulation of Na$_v$1.7 and Na$_v$1.8 peripheral nerve sodium channels by the local anesthetic lidocaine. Br J Pharmacol 2004; 142(3): 576-84.
[http://dx.doi.org/10.1038/sj.bjp.0705796] [PMID: 15148257]

[135] Chapman JB. Consistency between thermodynamics and the kinetics of n, m, and h in the Hodgkin-Huxley equations. J Theor Biol 1980; 85(3): 487-95.
[http://dx.doi.org/10.1016/0022-5193(80)90322-7] [PMID: 7442275]

[136] Chiu SY. Inactivation of sodium channels: second order kinetics in myelinated nerve. J Physiol 1977; 273(3): 573-96.
[http://dx.doi.org/10.1113/jphysiol.1977.sp012111] [PMID: 304888]

[137] Kniffki K-D, Siemen D, Vogel W. Development of sodium permeability inactivation in nodal membranes. J Physiol 1981; 313(313): 37-48.
[http://dx.doi.org/10.1113/jphysiol.1981.sp013649] [PMID: 7277227]

[138] Chandler WK, Meves H. Evidence for two types of sodium conductance in axons perfused with sodium fluoride solution. J Physiol 1970; 211(3): 653-78.
[http://dx.doi.org/10.1113/jphysiol.1970.sp009298] [PMID: 5501056]

[139] Sigg D, Bezanilla F. Total charge movement per channel. The relation between gating charge displacement and the voltage sensitivity of activation. J Gen Physiol 1997; 109(1): 27-39.
[http://dx.doi.org/10.1085/jgp.109.1.27] [PMID: 8997663]

[140] Greeff NG, Kühn FJ. Sodium load of *Xenopus* oocytes during high expression of rBIIA sodium channels reduces the ratio of ion permeability to gating charge. Biophys J 2000; 78(1): 84. [*a*.].

[141] Hirschberg B, Rovner A, Lieberman M, Patlak J. Transfer of twelve charges is needed to open skeletal muscle Na+ channels. J Gen Physiol 1995; 106(6): 1053-68.
[http://dx.doi.org/10.1085/jgp.106.6.1053] [PMID: 8786350]

[142] Ruben PC, Fleig A, Featherstone D, Starkus JG, Rayner MD. Effects of clamp rise-time on rat brain IIA sodium channels in *Xenopus* oocytes. J Neurosci Methods 1997; 73(2): 113-22.
[http://dx.doi.org/10.1016/S0165-0270(96)02216-9] [PMID: 9196281]

[143] Bekkers JM, Forster IC, Greeff NG. Gating current associated with inactivated states of the squid axon gating channel. Proc Natl Acad Sci USA 1990; 87(21): 8311-5.
[http://dx.doi.org/10.1073/pnas.87.21.8311] [PMID: 2172981]

[144] Conti F, Stühmer W. Quantal charge redistributions accompanying the structural transitions of sodium channels. Eur Biophys J 1989; 17(2): 53-9.
[http://dx.doi.org/10.1007/BF00257102] [PMID: 2548829]

[145] Chahine M, George AL Jr, Zhou M, *et al.* Sodium channel mutations in paramyotonia congenita uncouple inactivation from activation. Neuron 1994; 12(2): 281-94.
[http://dx.doi.org/10.1016/0896-6273(94)90271-2] [PMID: 8110459]

[146] Chen L-Q, Santarelli V, Horn R, Kallen RG. A unique role for the S4 segment of domain 4 in the inactivation of sodium channels. J Gen Physiol 1996; 108(6): 549-56.
[http://dx.doi.org/10.1085/jgp.108.6.549] [PMID: 8972392]

[147] Ji S, George AL Jr, Horn R, Barchi RL. *Paramyotonia congenita* mutations reveal different roles for segments S3 and S4 of domain D4 in hSkM1 sodium channel gating. J Gen Physiol 1996; 107(2): 183-94.
[http://dx.doi.org/10.1085/jgp.107.2.183] [PMID: 8833340]

[148] Kontis KJ, Goldin AL. Sodium channel inactivation is altered by substitution of voltage sensor positive charges. J Gen Physiol 1997; 110(4): 403-13.
[http://dx.doi.org/10.1085/jgp.110.4.403] [PMID: 9379172]

[149] Greeff NG, Kühn FJ. Variable ratio of permeability to gating charge of rBIIA sodium channels and sodium influx in *Xenopus* oocytes. Biophys J 2000; 79(5): 2434-53.
[http://dx.doi.org/10.1016/S0006-3495(00)76487-3] [PMID: 11053121]

[150] Perozo E, MacKinnon R, Bezanilla F, Stefani E. Gating currents from a nonconducting mutant reveal open-closed conformations in *Shaker* K+ channels. Neuron 1993; 11(2): 353-8.
[http://dx.doi.org/10.1016/0896-6273(93)90190-3] [PMID: 8352943]

[151] Perozo E, Santacruz-Toloza L, Stefani E, Bezanilla F, Papazian DM. S4 mutations alter gating currents of *Shaker* K channels. Biophys J 1994; 66(2 Pt 1): 345-54.
[http://dx.doi.org/10.1016/S0006-3495(94)80783-0] [PMID: 8161688]

[152] Bezanilla F, Perozo E, Stefani E. Gating of *Shaker* K+ channels: II. The components of gating currents and a model of channel activation. Biophys J 1994; 66(4): 1011-21.
[http://dx.doi.org/10.1016/S0006-3495(94)80882-3] [PMID: 8038375]

[153] Aggarwal SK, MacKinnon R. Contribution of the S4 segment to gating charge in the *Shaker* K+ channel. Neuron 1996; 16(6): 1169-77.
[http://dx.doi.org/10.1016/S0896-6273(00)80143-9] [PMID: 8663993]

[154] Long SB, Campbell EB, Mackinnon R. Voltage sensor of Kv1.2: structural basis of electromechanical coupling. Science 2005; 309(5736): 903-8.
[http://dx.doi.org/10.1126/science.1116270] [PMID: 16002579]

[155] Schoppa NE, McCormack K, Tanouye MA, Sigworth FJ. The size of gating charge in wild-type and mutant *Shaker* potassium channels. Science 1992; 255(5052): 1712-5.

[http://dx.doi.org/10.1126/science.1553560] [PMID: 1553560]

[156] Aggarwal SK, MacKinnon R. Contribution of the S4 segment to gating charge in the *Shaker* K+ channel. Neuron 1996; 16(6): 1169-77.
[http://dx.doi.org/10.1016/S0896-6273(00)80143-9] [PMID: 8663993]

[157] Seoh S-A, Sigg D, Papazian DM, Bezanilla F. Voltage-sensing residues in the S2 and S4 segments of the *Shaker* K+ channel. Neuron 1996; 16(6): 1159-67.
[http://dx.doi.org/10.1016/S0896-6273(00)80142-7] [PMID: 8663992]

[158] Li Y, Gao J, Lu Z, *et al.* Intracellular ATP binding is required to activate the slowly activating K+ channel *I*(Ks). Proc Natl Acad Sci USA 2013; 110(47): 18922-7.
[http://dx.doi.org/10.1073/pnas.1315649110] [PMID: 24190995]

[159] Zaydman MA, Cui J. PIP2 regulation of KCNQ channels: biophysical and molecular mechanisms for lipid modulation of voltage-dependent gating. Front Physiol 2014; 5: 195.
[http://dx.doi.org/10.3389/fphys.2014.00195] [PMID: 24904429]

[160] Li Y, Zaydman MA, Wu D, *et al.* KCNE1 enhances phosphatidylinositol 4,5-bisphosphate (PIP2) sensitivity of *I*Ks to modulate channel activity. Proc Natl Acad Sci USA 2011; 108(22): 9095-100.
[http://dx.doi.org/10.1073/pnas.1100872108] [PMID: 21576493]

[161] Zaydman MA, Silva JR, Delaloye K, *et al.* K$_v$7.1 ion channels require a lipid to couple voltage sensing to pore opening. Proc Natl Acad Sci USA 2013; 110(32): 13180-5.
[http://dx.doi.org/10.1073/pnas.1305167110] [PMID: 23861489]

[162] Marx SO, Kurokawa J, Reiken S, *et al.* Requirement of a macromolecular signaling complex for beta adrenergic receptor modulation of the KCNQ1-KCNE1 potassium channel. Science 2002; 295(5554): 496-9.
[http://dx.doi.org/10.1126/science.1066843] [PMID: 11799244]

[163] Jan LY, Jan YN. Receptor-regulated ion channels. Curr Opin Cell Biol 1997; 9(2): 155-60.
[http://dx.doi.org/10.1016/S0955-0674(97)80057-9] [PMID: 9069261]

[164] Vivas O, Arenas I, Garcia DE. Gating charge movement in native cells: another application of the patch clamp technique. In: Kaneez FD, Ed. Patch clamp technique. InTech 2012; pp. 255-66.
[http://dx.doi.org/10.5772/34899]

[165] Hille B. Modulation of ion-channel function by G-protein-coupled receptors. Trends Neurosci 1994; 17(12): 531-6.
[http://dx.doi.org/10.1016/0166-2236(94)90157-0] [PMID: 7532338]

[166] Carabelli V, Lovallo M, Magnelli V, Zucker H, Carbone E. Voltage-dependent modulation of single N-Type Ca2+ channel kinetics by receptor agonists in IMR32 cells. Biophys J 1996; 70(5): 2144-54.
[http://dx.doi.org/10.1016/S0006-3495(96)79780-1] [PMID: 9172738]

[167] Dunlap K, Fischbach GD. Neurotransmitters decrease the calcium conductance activated by depolarization of embryonic chick sensory neurones. J Physiol 1981; 317(317): 519-35.
[http://dx.doi.org/10.1113/jphysiol.1981.sp013841] [PMID: 6118434]

[168] Ma JY, Li M, Catterall WA, Scheuer T. Modulation of brain Na+ channels by a G-protein-coupled pathway. Proc Natl Acad Sci USA 1994; 91(25): 12351-5.
[http://dx.doi.org/10.1073/pnas.91.25.12351] [PMID: 7991631]

[169] Dolphin AC. G protein modulation of voltage-gated calcium channels. Pharmacol Rev 2003; 55(4): 607-27.
[http://dx.doi.org/10.1124/pr.55.4.3] [PMID: 14657419]

[170] Hernández-Ochoa EO, García-Ferreiro RE, García DE. G protein activation inhibits gating charge movement in rat sympathetic neurons. Am J Physiol Cell Physiol 2007; 292(6): C2226-38.
[http://dx.doi.org/10.1152/ajpcell.00540.2006] [PMID: 17314266]

[171]   Rebolledo-Antúnez S, Farías JM, Arenas I, García DE. Gating charges per channel of Ca(V)2.2 channels are modified by G protein activation in rat sympathetic neurons. Arch Biochem Biophys 2009; 486(1): 51-7.
[http://dx.doi.org/10.1016/j.abb.2009.04.002] [PMID: 19364492]

[172]   Noceti F, Baldelli P, Wei X, *et al.* Effective gating charges per channel in voltage-dependent K+ and Ca2+ channels. J Gen Physiol 1996; 108(3): 143-55.
[http://dx.doi.org/10.1085/jgp.108.3.143] [PMID: 8882860]

[173]   Crouzy SC, Sigworth FJ. Fluctuations in ion channel gating currents. Analysis of nonstationary shot noise. Biophys J 1993; 64(1): 68-76.
[http://dx.doi.org/10.1016/S0006-3495(93)81341-9] [PMID: 8381683]

[174]   Sigg D, Qian H, Bezanilla F. Kramers diffusion theory applied to gating kinetics of voltage-dependent ion channels. Biophys J 1999; 76(2): 782-803.
[http://dx.doi.org/10.1016/S0006-3495(99)77243-7] [PMID: 9929481]

[175]   Bezanilla F. The voltage sensor in voltage-dependent ion channels. Physiol Rev 2000; 80(2): 555-92.
[PMID: 10747201]

[176]   Sigg D, Stefani E, Bezanilla F. Gating current noise produced by elementary transitions in *Shaker* potassium channels. Science 1994; 264(5158): 578-82.
[http://dx.doi.org/10.1126/science.8160016] [PMID: 8160016]

[177]   Stefani E, Sigg D, Bezanilla F. Gating current fluctuations in normal and slow-inactivated *Shaker* K channels. Biophys J 1999; 76(1): A192. [Abstract].

[178]   Kostyuk E, Kostyuk P, Voitenko N. Structural and functional characteristics of nociceptive pathways and their alterations. Neurophysiology 2001; 33(4): 303-13.
[http://dx.doi.org/10.1023/A:1013532902247]

[179]   Akopian AN, Souslova V, England S, *et al.* The tetrodotoxin-resistant sodium channel SNS has a specialized function in pain pathways. Nat Neurosci 1999; 2(6): 541-8.
[http://dx.doi.org/10.1038/9195] [PMID: 10448219]

[180]   Akopian AN, Sivilotti L, Wood JN. A tetrodotoxin-resistant voltage-gated sodium channel expressed by sensory neurons. Nature 1996; 379(6562): 257-62.
[http://dx.doi.org/10.1038/379257a0] [PMID: 8538791]

[181]   Chambers JC, Zhao J, Terracciano CM, *et al.* Genetic variation in SCN10A influences cardiac conduction. Nat Genet 2010; 42(2): 149-52.
[http://dx.doi.org/10.1038/ng.516] [PMID: 20062061]

[182]   Damarjian TG, Craner MJ, Black JA, Waxman SG. Upregulation and colocalization of p75 and $Na_v1.8$ in Purkinje neurons in experimental autoimmune encephalomyelitis. Neurosci Lett 2004; 369(3): 186-90.
[http://dx.doi.org/10.1016/j.neulet.2004.07.023] [PMID: 15464262]

[183]   Han C, Huang J, Waxman SG. Sodium channel $Na_v1.8$: Emerging links to human disease. Neurology 2016; 86(5): 473-83.
[http://dx.doi.org/10.1212/WNL.0000000000002333] [PMID: 26747884]

[184]   Smith TH, Grider JR, Dewey WL, Akbarali HI. Morphine decreases enteric neuron excitability *via* inhibition of sodium channels. PLoS One 2012; 7(9): e45251.
[http://dx.doi.org/10.1371/journal.pone.0045251] [PMID: 23028881]

[185]   Gold MS. Tetrodotoxin-resistant Na+ currents and inflammatory hyperalgesia. Proc Natl Acad Sci USA 1999; 96(14): 7645-9.
[http://dx.doi.org/10.1073/pnas.96.14.7645] [PMID: 10393874]

[186]   Black JA, Liu S, Tanaka M, Cummins TR, Waxman SG. Changes in the expression of tetrodotoxin-sensitive sodium channels within dorsal root ganglia neurons in inflammatory pain. Pain 2004; 108(3):

237-47.
[http://dx.doi.org/10.1016/j.pain.2003.12.035] [PMID: 15030943]

[187] Gould HJ III, England JD, Soignier RD, *et al.* Ibuprofen blocks changes in Na$_v$1.7 and 1.8 sodium channels associated with complete Freunds adjuvant-induced inflammation in rat. J Pain 2004; 5(5): 270-80.
[http://dx.doi.org/10.1016/j.jpain.2004.04.005] [PMID: 15219259]

[188] Fischer MJ, Mak SW, McNaughton PA. Sensitization of nociceptors – what are ion channels doing? Open Pain J 2010; 3: 82-96.
[http://dx.doi.org/10.2174/1876386301003010082]

[189] Read HL, Siegel RM. The origins of aperiodicities in sensory neuron entrainment. Neuroscience 1996; 75(1): 301-14.
[http://dx.doi.org/10.1016/0306-4522(96)00227-8] [PMID: 8923543]

[190] England S, Bevan S, Docherty RJ. PGE2 modulates the tetrodotoxin-resistant sodium current in neonatal rat dorsal root ganglion neurones *via* the cyclic AMP-protein kinase A cascade. J Physiol 1996; 495(Pt 2): 429-40.
[http://dx.doi.org/10.1113/jphysiol.1996.sp021604] [PMID: 8887754]

[191] Han C, Estacion M, Huang J, *et al.* Human Na$_{(v)}$1.8: enhanced persistent and ramp currents contribute to distinct firing properties of human DRG neurons. J Neurophysiol 2015; 113(9): 3172-85.
[http://dx.doi.org/10.1152/jn.00113.2015] [PMID: 25787950]

[192] Keh SM, Facer P, Simpson KD, Sandhu G, Saleh HA, Anand P. Increased nerve fiber expression of sensory sodium channels Na$_v$1.7, Na$_v$1.8, and Na$_v$1.9 in rhinitis. Laryngoscope 2008; 118(4): 573-9.
[http://dx.doi.org/10.1097/MLG.0b013e3181625d5a] [PMID: 18197135]

[193] Shields SD, Cheng X, Gasser A, *et al.* A channelopathy contributes to cerebellar dysfunction in a model of multiple sclerosis. Ann Neurol 2012; 71(2): 186-94.
[http://dx.doi.org/10.1002/ana.22665] [PMID: 22367990]

[194] Pert CB, Snyder SH. Opiate receptor: demonstration in nervous tissue. Science 1973; 179(4077): 1011-4.
[http://dx.doi.org/10.1126/science.179.4077.1011] [PMID: 4687585]

[195] Simon EJ, Hiller JM, Edelman I. Stereospecific binding of the potent narcotic analgesic [3H]etorphine to rat-brain homogenate. Proc Natl Acad Sci USA 1973; 70(7): 1947-9.
[http://dx.doi.org/10.1073/pnas.70.7.1947] [PMID: 4516196]

[196] Terenius L. Stereospecific interaction between narcotic analgesics and a synaptic plasma membrane fraction of rat cerebral cortex. Acta Pharmacol Toxicol (Copenh) 1973; 32(3): 317-20.
[PMID: 4801733]

[197] Mollereau C, Parmentier M, Mailleux P, *et al.* ORL1, a novel member of the opioid receptor family. Cloning, functional expression and localization. FEBS Lett 1994; 341(1): 33-8.
[http://dx.doi.org/10.1016/0014-5793(94)80235-1] [PMID: 8137918]

[198] Frost JJ, Wagner HN Jr, Dannals RF, *et al.* Imaging opiate receptors in the human brain by positron tomography. J Comput Assist Tomogr 1985; 9(2): 231-6.
[http://dx.doi.org/10.1097/00004728-198503000-00001] [PMID: 2982931]

[199] Martin WR. Opioid antagonists. Pharmacol Rev 1967; 19(4): 463-521.
[PMID: 4867058]

[200] Bodnar RJ. Endogenous opiates and behavior: 2012. Peptides 2013; 50: 55-95.
[http://dx.doi.org/10.1016/j.peptides.2013.10.001] [PMID: 24126281]

[201] Jessell TM, Iversen LL. Opiate analgesics inhibit substance P release from rat trigeminal nucleus. Nature 1977; 268(5620): 549-51.
[http://dx.doi.org/10.1038/268549a0] [PMID: 18681]

[202] Macdonald RL, Nelson PG. Specific-opiate-induced depression of transmitter release from dorsal root ganglion cells in culture. Science 1978; 199(4336): 1449-51.
[http://dx.doi.org/10.1126/science.204015] [PMID: 204015]

[203] Mudge AW, Leeman SE, Fischbach GD. Enkephalin inhibits release of substance P from sensory neurons in culture and decreases action potential duration. Proc Natl Acad Sci USA 1979; 76(1): 526-30.
[http://dx.doi.org/10.1073/pnas.76.1.526] [PMID: 218204]

[204] Grudt TJ, Williams JT. μ-Opioid agonists inhibit spinal trigeminal substantia gelatinosa neurons in guinea pig and rat. J Neurosci 1994; 14(3 Pt 2): 1646-54.
[PMID: 8126561]

[205] Ippolito DL, Temkin PA, Rogalski SL, Chavkin C. N-terminal tyrosine residues within the potassium channel Kir3 modulate GTPase activity of Galphai. J Biol Chem 2002; 277(36): 32692-6.
[http://dx.doi.org/10.1074/jbc.M204407200] [PMID: 12082117]

[206] Sadja R, Alagem N, Reuveny E. Gating of GIRK channels: details of an intricate, membrane-delimited signaling complex. Neuron 2003; 39(1): 9-12.
[http://dx.doi.org/10.1016/S0896-6273(03)00402-1] [PMID: 12848928]

[207] Rusin KI, Giovannucci DR, Stuenkel EL, Moises HC. κ-opioid receptor activation modulates Ca^{2+} currents and secretion in isolated neuroendocrine nerve terminals. J Neurosci 1997; 17(17): 6565-74.
[PMID: 9254669]

[208] Bourinet E, Soong TW, Stea A, Snutch TP. Determinants of the G protein-dependent opioid modulation of neuronal calcium channels. Proc Natl Acad Sci USA 1996; 93(4): 1486-91.
[http://dx.doi.org/10.1073/pnas.93.4.1486] [PMID: 8643659]

[209] Zamponi GW, Snutch TP. Modulating modulation: crosstalk between regulatory pathways of presynaptic calcium channels. Mol Interv 2002; 2(8): 476-8.
[http://dx.doi.org/10.1124/mi.2.8.476] [PMID: 14993397]

[210] Zamponi GW, Snutch TP. Modulation of voltage-dependent calcium channels by G proteins. Curr Opin Neurobiol 1998; 8(3): 351-6.
[http://dx.doi.org/10.1016/S0959-4388(98)80060-3] [PMID: 9687363]

[211] Krylov BV, Derbenev AV, Podzorova SA, Liudyno MI, Kuzmin AV, Izvarina NL. [Morphine decreases the voltage sensitivity of the slow sodium channels]. Ross Fiziol Zh Im I M Sechenova 1999; 85(2): 225-36.
[PMID: 10389179]

[212] Wolozin BL, Pasternak GW. Classification of multiple morphine and enkephalin binding sites in the central nervous system. Proc Natl Acad Sci USA 1981; 78(10): 6181-5.
[http://dx.doi.org/10.1073/pnas.78.10.6181] [PMID: 6273857]

[213] Osipchuk YV, Timin EN. Electrical measurements on perfused cells. In: Kostyuk PG, Kryshtal OA, Eds. Intracellular perfusion of excitable cells. New York: Wiley 1984; pp. 103-29.

CHAPTER 2

Possible Mechanisms of Binding of Gamma-Pyrones to the Opioid-Like Receptor

Abstract: Derivatives of gamma-pyrone show their remarkable ability to trigger the novel mechanism of $Na_V1.8$ channels modulation described in Chapter **1**. Unlike morphine, which activates both opioid and opioid-like receptors, comenic acid specifically switches on the latter mechanism involving Na^+,K^+-ATPase as the signal transducer. It is extremely important that not any gamma-pyrone derivative can decrease the voltage sensitivity of $Na_V1.8$ channels, though all molecules studied herein share a rather similar chemical structure. A very productive approach which makes it possible to elucidate the peculiarities of ligand-receptor binding on the molecular level is combined application of quantum-chemical calculations and the patch-clamp method. Below we present our findings that explain a totally unevident result of highly selective binding of gamma-pyrone derivatives to the opioid-like receptor. Understanding of this mechanism opens up opportunities for creation of a novel class of analgesics.

Keywords: Ca^{2+} chelate complex, Gamma-pyrone derivatives, Limiting slope procedure, $Na_V1.8$ channels, Nociception, Opioid-like receptor, Patch-clamp method, Quantum-chemical calculations.

Pharmacological effects of gamma-pyrone derivatives, including radioprotective [1 - 3], antiviral [4], antidiabetic [5], and anticonvulsant [6] effects, were recently examined. Kojic acid was demonstrated to be able to protect human skin from pigmentation [7, 8]. Gamma-pyrones are regarded as potential anticancer drugs [9]. They also exhibit antileishmanial activity [10].

Kojic acid derivatives were found to effectively modulate histamine H3 receptors (H3R). The most affine compounds showed receptor binding in the low nanomolar concentration range [11]. The authors suggest that antagonists/inverse agonists of the H3R are able to increase the neurotransmitter content and may find their application in the therapy of cognitive diseases, sleep/wake disorders, epilepsy, obesity, pain, or allergic rhinitis. Several substances are progressing in clinical trials [11].

Boris V. Krylov, Ilia V. Rogachevskii, Tatiana N. Shelykh, Vera B. Plakhova

Fundamentally new opportunities for clinical use of gamma-pyrones were discovered by Alexey Shurygin who developed food additive Baliz-2 [12, 13]. Its main ingredients, comenic and meconic acids, exhibit a profound antibiotic, antibacterial, and regenerative activity. It is worth noting that Baliz-2 never expressed any negative side effects during its long history of clinical application in Russia. Our starting investigations of probable molecular mechanisms of meconic and comenic acid targeting were inspiring: the agents decreased voltage sensitivity of $Na_V1.8$ channels [14]. These findings opened a promising perspective for research of the role of gamma-pyrone derivatives in nociception (see also [15]).

PATCH-CLAMP INVESTIGATION OF GAMMA-PYRONES

The families of $Na_V1.8$ currents in the control experiment and after extracellular application of comenic acid (5-hydroxy-gamma-pyrone-2-carboxylic acid, substance A) are presented in Fig. (**2.1.a**). It is clearly seen that the amplitude values of the currents are decreased, which can find its partial explanation in the existence of "run-down" effect inherent to the patch-clamp method [16, 17]. However, the decrease of the channels density may also take place. The peak current-voltage curve shifts in the depolarizing direction after comenic acid has been applied (Fig. **2.1.b**). The left branch of the current-voltage function is steeper as a result of comenic acid application than in the control experiments. The voltage dependencies of normalized $G_{Na\_s}(E)$ functions also differ between the control and comenic acid data at negative E (Fig. **2.2.a**). When $G_{Na\_s}(E)$ dependencies are obtained, the Almers' limiting slope procedure can be applied, making it possible to evaluate Z_{eff} at the most negative potentials E (Fig. **2.2.b**). A very pronounced decrease in Z_{eff} after extracellular application of comenic acid occurs due to activation of the receptor-coupled membrane mechanism (Fig. **1.17**). Indeed, a nonspecific opioid antagonist naltrexone (NTX) switched off the effect of comenic acid (Fig. **2.3**). Z_{eff} also remained fairly unchanged after combined application of comenic acid and a specific blocker of Na^+,K^+-ATPase, ouabain, at 200 μM (Fig. **2.3**). Ouabain applied at this rather high concentration totally inhibits Na^+,K^+-ATPase, therefore interrupting transduction of the signal triggered by binding of comenic acid to the opioid-like receptor and sent to $Na_V1.8$ channels according to the mechanism proposed earlier [18]. Moreover, these findings indicate that comenic acid can be compared to morphine in its efficiency of $Na_V1.8$ channel modulation. It switches on the three background mechanisms: reduces the channels density, positively shifts $Na_V1.8$ channel activation gating process, and, most importantly, markedly decreases Z_{eff}. The latter process is of dose-dependent nature, showing opioid-like receptor binding in the nanomolar concentration range. The binding process is characterized by $K_d =$ 100 nM and the Hill coefficient $n = 0.5$ [14].

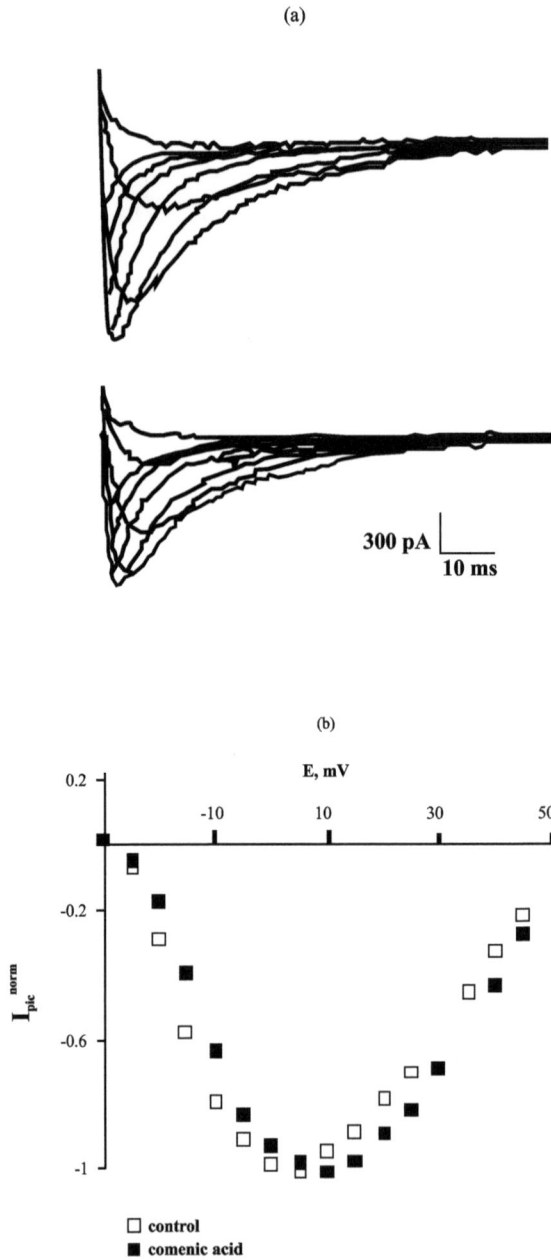

(a)

300 pA

10 ms

(b)

Fig. (2.1). Effects of comenic acid on $Na_v1.8$ channels.

a – Families of sodium currents measured in the control experiment (top) and after application of comenic acid at 100 nM (bottom);

b – Positive shift of the normalized peak current-voltage curve after application of comenic acid.

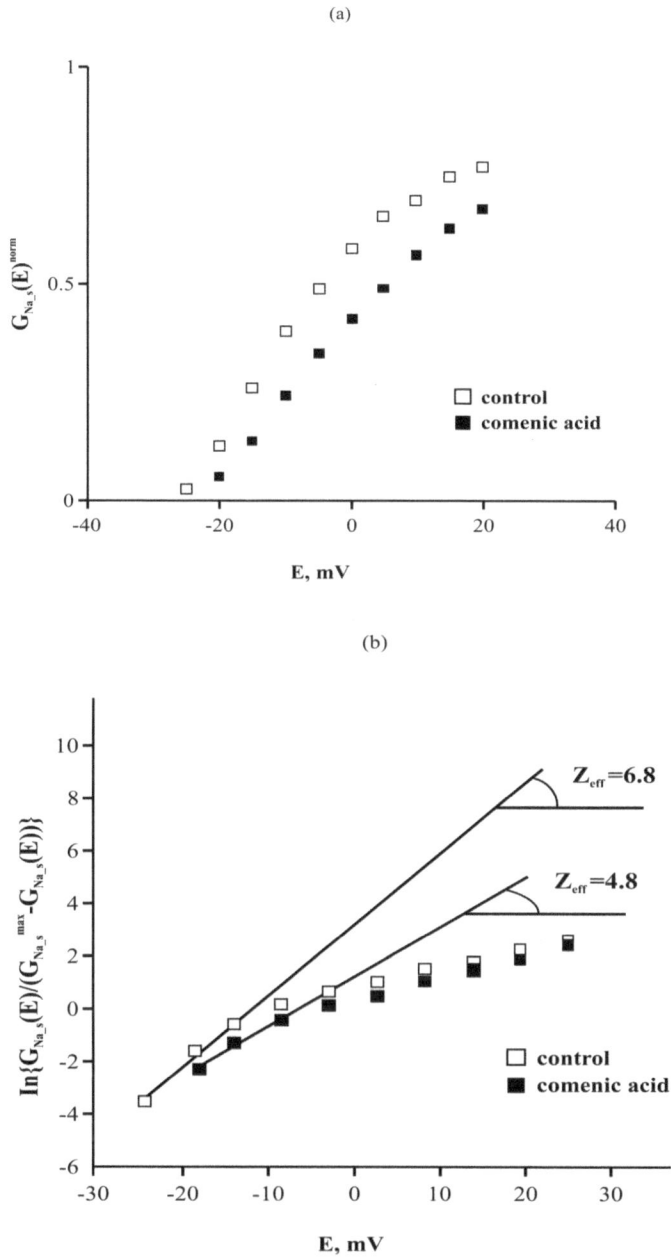

Fig. (2.2). Decrease of effective charge of $Na_V1.8$ channels activation gating device after application of comenic acid.

a – Voltage dependence of the normalized peak conductance used for Z_{eff} evaluation;

b – Z_{eff} evaluation by the Almers' limiting slope procedure in the control experiment and after application of comenic acid at 100 nM.

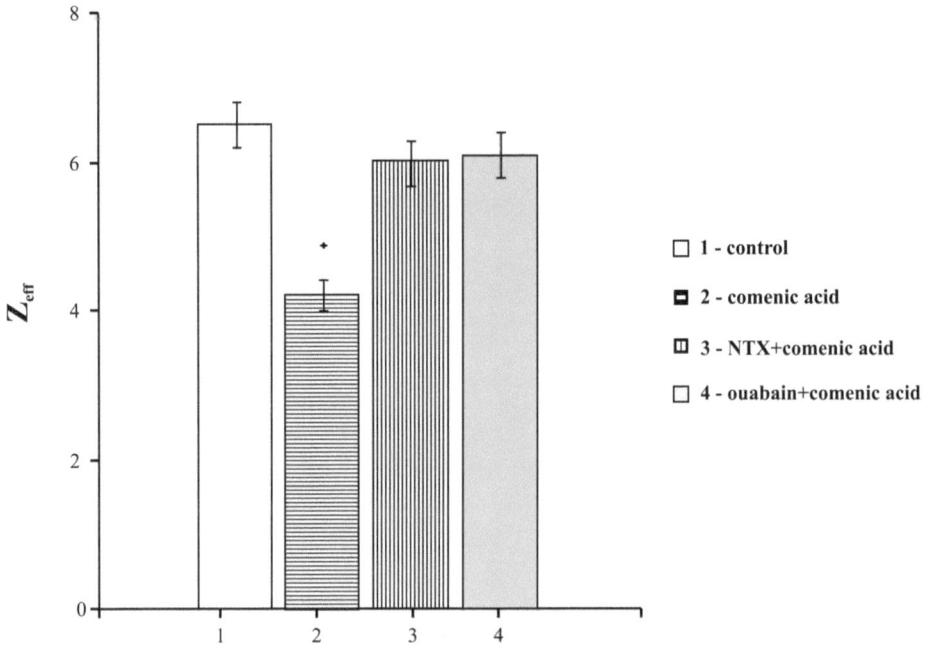

Fig. (2.3). Decrease of $Na_V1.8$ channels effective charge resulting from application of comenic acid is blocked by naltrexone (NTX) and ouabain.

The control value of Z_{eff} equals to 6.5±0.3 (n = 22). Z_{eff} is reduced to 4.2±0.2 (n = 18) after application of comenic acid at 1 μM. Combined application of NTX (50 μM) and comenic acid did not result in a statistically significant decrease of the effective charge (6.0±0.3, n = 12), as well as combined application of ouabain (200 μM) and comenic acid (6.1±0.3, n = 15).

* – difference between experimental and control data is statistically significant (only in the case of comenic acid application).

Our experimental results also demonstrate that one more gamma-pyrone derivative, in addition to comenic and meconic acids, is able to decrease Z_{eff} of $Na_V1.8$ channel activation gating system. It is 5-methoxy-gamma-pyrone-2-carboxylic acid (substance B). Its action is characterized by the same three manifestations of inhibitory properties: it evokes a decrease of the amplitude values of the currents and induces a positive voltage shift of the current-voltage relationship (Fig. **2.4**). The value of Z_{eff} is also strongly decreased after application of substance B (Fig. **2.5**). Substance B activates the same membrane signaling mechanism that includes the opioid-like receptor and Na^+,K^+-ATPase as the signal transducer. This conclusion is based on the fact that both naltrexone and ouabain totally block the effects of this substance (Fig. **2.6**).

(a)

(b)

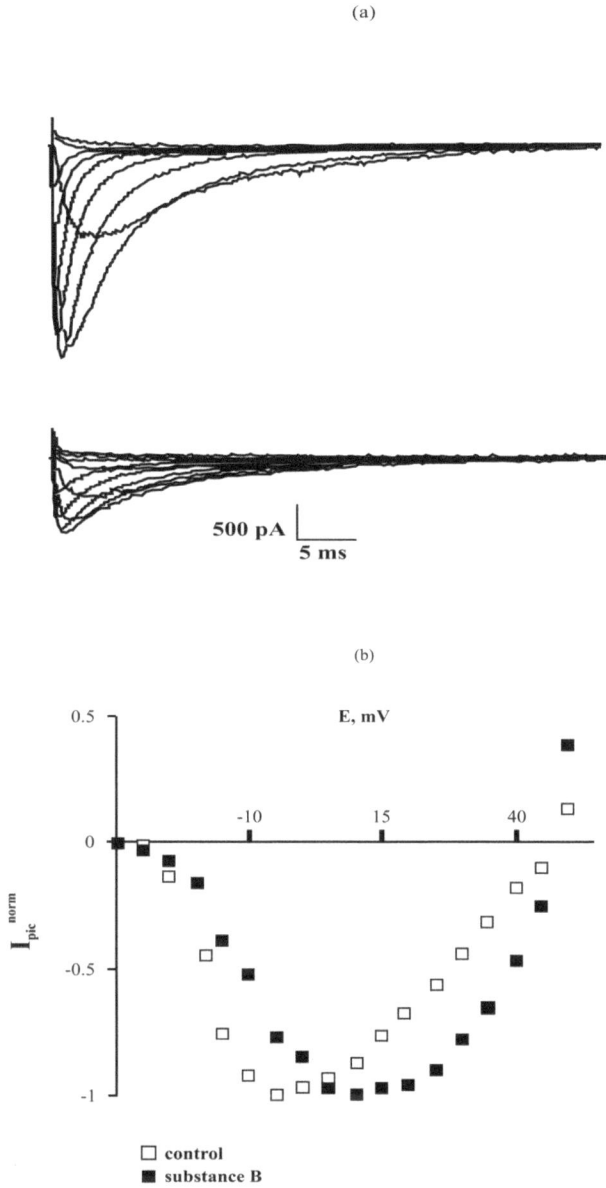

Fig. (2.4). Effects of substance B (5-methoxy-gamma-pyrone-2-carboxylic acid) on $Na_V 1.8$ channels.
a – Families of sodium currents measured in the control experiment (top) and after application of substance B at 100 nM (bottom);
b – Positive shift of the normalized peak current-voltage curve after application of substance B.

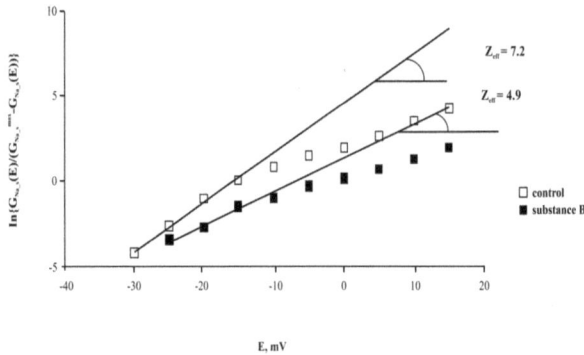

Fig. (2.5). Decrease of $Na_V1.8$ channels effective charge after application of substance B. Z_{eff} evaluation by the Almers' limiting slope procedure in the control experiment and after application of substance B at 100 nM.

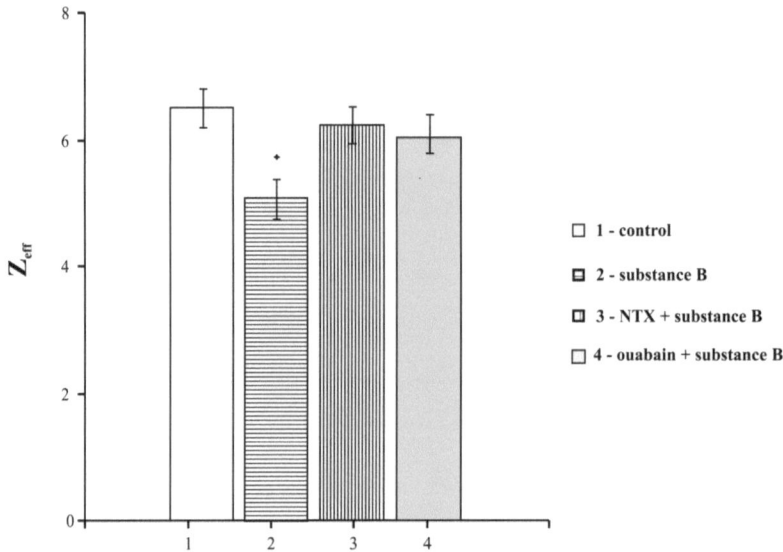

Fig. (2.6). Decrease of $Na_V1.8$ channels effective charge resulting from application of substance B is blocked by naltrexone (NTX) and ouabain.

Z_{eff} is reduced to 5.1±0.3 (n = 20) from the control value (6.5± 0.3, n = 22) after application of substance B at 100 nM. Combined application of NTX (50 μM) and substance B did not result in a statistically significant decrease of the effective charge (6.2±0.3, n = 18), as well as combined application of ouabain (200 μM) and substance B (6.0±0.4, n = 17).

* – difference between experimental and control data is statistically significant (only in the case of substance B application).

QUANTUM-CHEMICAL STUDY OF GAMMA-PYRONES

It is particularly important that not any gamma-pyrone derivative can decrease the voltage sensitivity of $Na_V1.8$ channels, though all molecules studied herein share a very similar chemical structure. Comenic acid and substance B pronouncedly change Z_{eff}, while chelidonic acid (gamma-pyrone-2,6-dicarboxylic acid, substance C) and kojic acid (5-hydroxy-2-hydroxymethyl-gamma-pyrone, substance D) do not (Fig. **2.7**). Structural formulae of substances A-D are presented in Fig. (**2.8**).

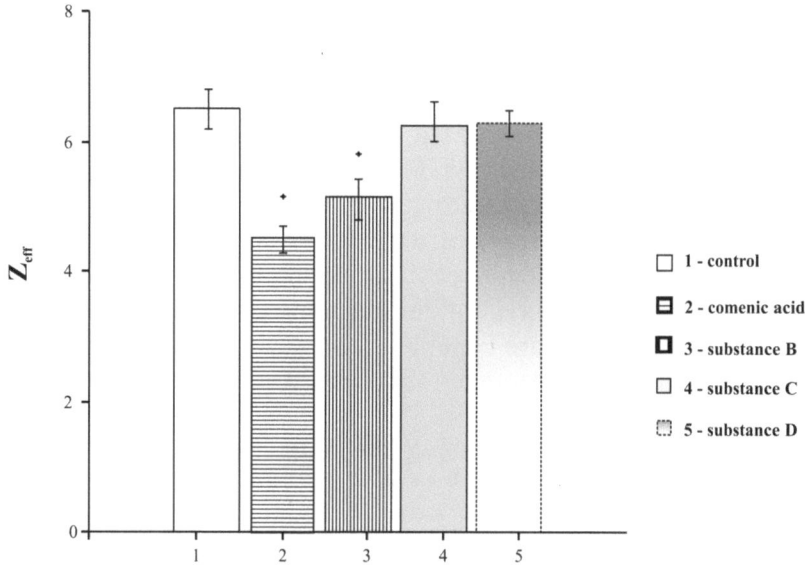

Fig. (2.7). Effects of gamma-pyrones on $Na_V 1.8$ channels: substance C (chelidonic acid) and substance D (kojic acid) do not decrease Z_{eff}.
The control value of Z_{eff} equals to 6.5±0.3 (n = 22). All gamma-pyrones were applied at 100 nM. Z_{eff} after application of comenic acid was equal to 4.5±0.2 (n = 17); substance B, 5.1±0.3 (n = 20); substance C, 6.2±0.4 (n = 20); substance D, 6.1±0.3 (n = 21).
* – difference between experimental and control data is statistically significant (in the cases of application of comenic acid and substance B.

A remarkably productive approach which makes it possible to elucidate the peculiarities of ligand-receptor binding on the molecular level is combined application of quantum-chemical calculations and the patch-clamp method. Below we present our findings that explain a totally unevident result of highly selective binding of gamma-pyrone derivatives to the opioid-like receptor. Understanding of this mechanism opens up opportunities for creation of a novel class of analgesics.

The structure of the opioid-like receptor remains so far unknown, so we have to base the logic of our further considerations on common structural features of opioid receptors and ligands described in the literature.

As a rule, opioid ligands contain at least one electrophilic moiety which may interact with negatively charged residues of their receptor upon binding [19]. Computer simulation and X-ray analysis made it possible to construct models of various opioid receptors (μ, δ, κ) [20 - 22] and thus give scientific credence to a tremendous amount of experimental data regarding affinities of diverse ligands, structure-activity relationship in series of closely related ligands and effects of mutagenesis on receptor properties and functioning. Of particular importance is the fact that two aspartate residues, absolutely conserved in all described types of opioid receptors, play an important role in binding of opioid ligands. It is not unreasonable to assume that the opioid-like receptor coupled to $Na_V1.8$ channels should also contain the specified residues. As neither comenic acid nor substance B includes any positively charged groups, it also makes sense to suggest that they bind to the receptor in the form of a complex with one or several inorganic cations. The carboxyl groups in α position to the pyrone ring oxygen atom can form salts, whereas the hydroxyl and the carbonyl groups which occupy adjacent positions of the heterocycle are capable of chelating di- and trivalent cations in aqueous solutions and water-dioxane mixtures [23 - 25].

Possible involvement of inorganic cations in ligand-receptor binding was intensively studied by Boris Zhorov and Vettai Ananthanarayanan, as the researchers tried to explain the ability of structurally diverse molecules to bind to the μ-opioid receptor. It was shown [22, 26 - 28] that in some cases μ-opioid ligands could effectively chelate small inorganic cations, such as Na^+ and Ca^{2+}, in order to 1) adopt a required for binding conformation, usually more compact and complementary to receptor binding pocket, and 2) introduce a positively charged moiety capable to form energetically favorable ion-ionic bonds with the receptor.

To find out which functional groups of gamma-pyrones that might interact with inorganic cations determine the ability of these molecules to modulate the activity of $Na_V1.8$ channels, two other gamma-pyrone derivatives were also examined in patch-clamp experiments: chelidonic and kojic acids (Fig. **2.7**). Chelidonic acid lacks a nucleophilic substituent next to its carbonyl group, and it is thus incapable of chelating cations, while being able to form salts. Kojic acid has no carboxyl group and cannot form salts, but this molecule can chelate cations. Substance B contains the methoxy group instead of the hydroxyl group present in comenic acid, and this should result in a lesser stability of chelate complexes involving the methoxy and carbonyl groups of substance B, as compared to chelates formed by comenic acid.

In our patch-clamp experiments, all agents studied were added to the extracellular solution containing Na^+, Ca^{2+} and Mg^{2+} at 65, 2 and 2 mM, correspondingly [18]. Various divalent cations Mt^{2+} (Mt = Ca, Cd, Co, Ni, Zn, Cu) were demonstrated to be effectively chelated by comenic acid and kojic acid in 1:1 stoichiometry at respective concentrations of cation and chelating agent equal to 100 μM and 1-2 mM [23 - 25]. Such chelate complexes are stable in solutions with ionic strengths of 0-1.5 M [24]. The ionic strength was modified by addition of electrolytes containing monovalent cations [24], therefore the presence of Na^+ in high concentrations should not affect the stability of chelates of divalent cations with gamma-pyrone derivatives under consideration. Chelate complexes of kojic acid with Ca^{2+} and Mg^{2+} have almost the same stability constants in aqueous solutions [25]. Experimental extracellular solution contained these cations in equal concentrations, several orders of magnitude higher than those of the agents studied. It may be thus accepted that gamma-pyrones formed 1:1 chelate complexes with divalent cations. The ionic compositions of the solutions used in our *in vitro* studies and, in particular, the ratio of Ca^{2+} to Mg^{2+} concentrations were selected so as to provide the maximal stability of cellular membranes of isolated neurons, and the solutions used should be regarded only as models of actual physiological conditions. Mg^{2+} is known to be a nonspecific blocker of synaptic transmission, and this quite negative physiological effect is already manifested at fairly low Mg^{2+} concentrations (3 mM) [29]. Hence, we will consider only Ca^{2+} chelation, even more so as Ca^{2+} is the cation that was demonstrated to be involved in binding of opioid ligands [26 - 28].

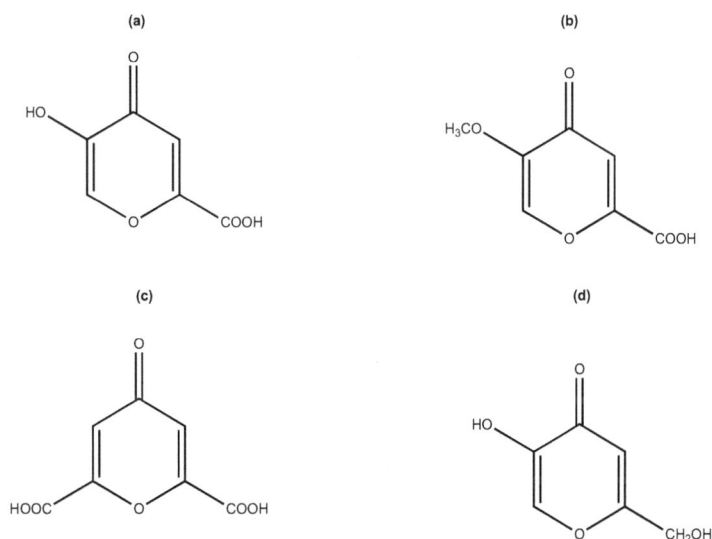

Fig. (2.8). Structural formulae of gamma-pyrones.
a – comenic acid (substance A);
b – 5-methoxy-gamma-pyrone-2-carboxylic acid (substance B);
c – chelidonic acid (substance C);
d – kojic acid (substance D).

In order to explain the results of patch-clamp experiments on the molecular level, a quantum-chemical study of gamma-pyrones in various molecular forms was carried out to investigate their ability to interact with inorganic cations and to propose a mechanism of their ligand-receptor binding. Full *ab initio* geometry optimization of studied gamma-pyrones in forms of free acids, their anions, sodium and calcium salts, Ca^{2+} chelate complexes, as well as sodium and calcium salts of the chelate complexes was performed by RHF method with 6-31G* basis set [30]. Calculations were made using the GAMESS package [31], mostly in the gas phase approximation (dielectric constant $\varepsilon = 1$). However, solvation effects were taken into account for comenic acid and its derivatives in the framework of the polarizable continuum model (PCM) [32] with $\varepsilon = 78.3$ (water solution) and $\varepsilon = 10$ (simulation of dielectric properties of receptor binding pocket). Steric structures of the studied agents in the forms including the maximal amount of bound cations and the numbering of atoms are presented in Fig. (**2.9**). The results of our calculations will be further divided into subgroups according to molecular forms of examined gamma-pyrones.

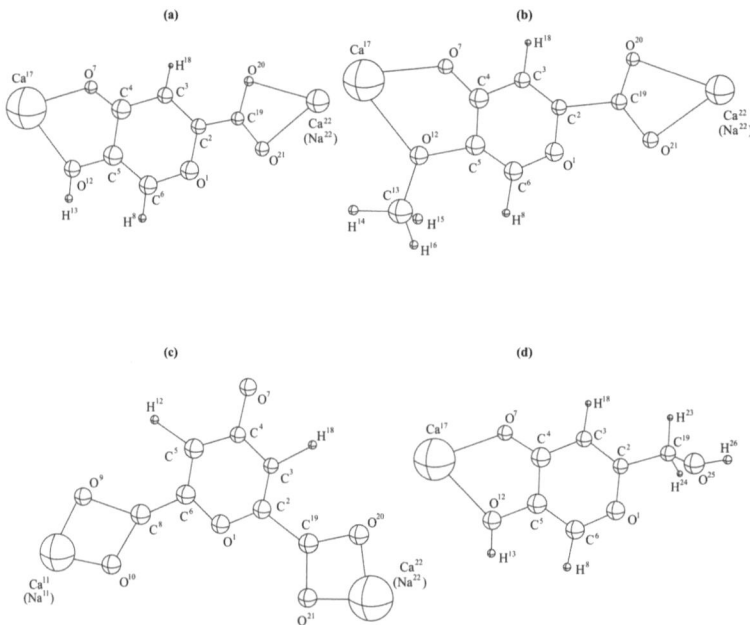

Fig. (2.9). Steric structures of gamma-pyrones in the forms including the maximal amount of bound cations and accepted atom numbering.
a – calcium (sodium) salt of calcium chelate complex of comenic acid (substance A);
b – calcium (sodium) salt of calcium chelate complex of 5-methoxy-gamma-pyrone-2-carboxylic acid (substance B);
c – calcium (sodium) salt of chelidonic acid (substance C);
d – calcium chelate complex of kojic acid (substance D).

Free Acids

Selected bond lengths, bond angles and atomic charges in the molecules of free acids A-D are presented in Table **2.1**. All molecules have a planar structure due to the presence of conjugated double bonds C^2—C^3, C^4—O^7, C^5—C^6, C^8—O^9 and C^{19}—O^{20}. Only the hydroxymethyl group in D ($O^1C^2C^{19}O^{25}$ torsion angle τ = -63.4°) and the methoxy group in B [$\tau(C^4C^5O^{12}C^{13})$ = -66.3°] deviate from the pyrone ring plane. The differences in the geometry of molecules A-D and in the charge distribution are insignificant. They are caused by the nature of ring substituents and presence of intramolecular hydrogen bonds. Here and hereinafter, parameters of an isolated molecule A will be used as a reference.

Molecules A and D contain weak hydrogen bonds O^{12}—$H^{13}\cdots O^7$, $r(O^7\cdots H^{13})$ distances being 2.167 and 2.152 Å, respectively. Replacement of the carboxyl group with the hydroxymethyl group in position 2 of the pyrone ring results in a decrease of $O^1C^2C^3$ and $O^1C^2C^{19}$ angles and charge redistribution on C^2 and C^3 atoms in molecule D. In molecules B and C, intramolecular hydrogen bonds are not formed, which accounts for elongation of C^2—O^1, C^3—C^4, and C^5—O^{12} bonds, strengthening of C^4—O^7 and C^6—O^1 bonds, increase of $C^4C^5O^{12}$, $C^5C^4O^7$, and $C^5C^6O^1$ bond angles, and decrease of the absolute values of O^7 and O^{12} atomic charges. The carboxyl group in position 6 of the pyrone ring in molecule C induces changes in $q(C^5)$ and $q(C^6)$.

Table 2.1. Selected bond lengths, bond angles, and atomic charges in the free acids A-D obtained by full 6-31G*/RHF geometry optimization with varied dielectric constant ε.

Parameter	A			B	C	D
	$\varepsilon = 1$	$\varepsilon = 10$	$\varepsilon = 78.3$	$\varepsilon = 1$	$\varepsilon = 1$	$\varepsilon = 1$
Bond Length, r, Å						
C^2—O^1	1.328	1.327	1.327	1.336	1.341	1.329
C^2—C^3	1.333	1.333	1.333	1.328	1.328	1.336
C^3—C^4	1.458	1.455	1.454	1.471	1.472	1.452
C^4—C^5	1.468	1.463	1.462	1.474	1.472	1.469
C^5—C^6	1.328	1.330	1.330	1.331	1.328	1.326
C^6—O^1	1.350	1.347	1.346	1.343	1.341	1.352
C^4—O^7	1.204	1.210	1.211	1.199	1.196	1.206
C^2—C^{19}	1.499	1.500	1.501	1.497	1.498	1.501
C^{19}—O^{20}	1.185	1.189	1.189	1.185	1.185	
C^{19}—O^{21}	1.317	1.311	1.310	1.318	1.317	
C^5—O^{12}	1.339	1.344	1.345	1.347		1.341

(Table 2.1) contd.....

Parameter	A			B	C	D
	$\varepsilon = 1$	$\varepsilon = 10$	$\varepsilon = 78.3$	$\varepsilon = 1$	$\varepsilon = 1$	$\varepsilon = 1$
C^6—C^8					1.498	
C^8—O^9					1.185	
C^8—O^{10}					1.317	
Bond Angle, ω, deg						
$O^1C^2C^3$	124.1	123.8	123.8	123.7	124.0	122.9
$C^2C^3C^4$	119.5	119.6	119.6	120.7	120.2	120.3
$C^3C^4C^5$	114.0	114.1	114.2	112.9	113.1	114.0
$C^4C^5C^6$	120.4	120.4	120.4	119.9	120.2	120.2
$C^5C^6O^1$	122.5	122.4	122.4	123.2	124.0	122.4
$C^6O^1C^2$	119.4	119.6	119.7	118.6	118.5	120.2
$C^4C^5O^{12}$	117.1	117.7	117.7	120.8		117.0
$C^5C^4O^7$	120.4	120.8	120.8	123.7	123.5	120.0
$O^1C^2C^{19}$	114.5	114.4	114.4	120.7	114.2	112.2
$C^2C^{19}O^{21}$	113.2	113.0	113.1	112.9	113.2	
$O^{20}C^{19}O^{21}$	124.3	124.8	124.8	119.9	124.3	
$O^1C^6C^8$					114.2	
$C^6C^8O^{10}$					113.2	
$O^9C^8O^{10}$					124.3	
Atomic Charge, q, a.u.						
O^1	-0.61	-0.61	-0.61	-0.61	-0.64	-0.61
C^2	0.32	0.34	0.34	0.31	0.31	0.45
C^3	-0.32	-0.32	-0.32	-0.32	-0.31	-0.39
C^4	0.54	0.55	0.56	0.55	0.56	0.55
C^5	0.26	0.25	0.25	0.25	-0.31	0.25
C^6	0.11	0.12	0.13	0.13	0.31	0.11
O^7	-0.61	-0.66	-0.67	-0.58	-0.56	-0.63
C^8					0.77	
O^9					-0.54	
O^{10}					-0.68	
O^{12}	-0.76	-0.78	-0.78	-0.65		-0.76
C^{19}	0.77	0.80	0.81	0.77	0.77	-0.03
O^{20}	-0.54	-0.58	-0.58	-0.54	-0.54	
O^{21}	-0.68	-0.69	-0.69	-0.68	-0.68	

Variation of the dielectric constant ε value (in other words, taking the solvation effects into account) barely changes the geometry of molecule A, but increases the negative charges on O^7, O^{12}, and O^{20} atoms, which invokes elongation of O^{12}—H^{13}···O^7 hydrogen bond ($r(O^7···H^{13})$ is equal to 2.202 and 2.205 Å at $\varepsilon = 10$ and $\varepsilon = 78.3$, respectively), C^4—O^7 and C^5—O^{12} bonds.

Acid Anions

The acidity constants of A, C, and D are as follows: $pK_a(COOH) = 1.4$ [33], $pK_a(OH) = 7.8$ [34] (A); $pK_{a1}(COOH) \approx pK_{a2}(COOH) = 2.36$ [35] (C); $pK_a(OH) = 7.91$ [36] (D). As our patch-clamp data were obtained under physiological conditions (pH = 7.2), only the carboxyl groups were considered ionized. Thus, the anions of A and B were singly charged; the anion of C, doubly charged; and molecule D was not examined in the anionic form.

All anion molecules are planar, except for the methoxy group in the anion of B [$\tau(C^4C^5O^{12}C^{13}) = -66.7°$]. The differences in steric and electronic structure in the series of anions of A-C are mostly the same as those in the corresponding free acids, which is illustrated by Table **2.2**. The effect of ε on the structure of the carboxyl group in the anion of A is noticeable. The growth of negative charge on O^{20} and O^{21} atoms results in elongation of C^{19}—O^{20}, C^{19}—O^{21} bonds and strengthening of C^2—C^{19} bond.

Table 2.2. Selected bond lengths, bond angles, and atomic charges in the anions of A-C obtained by full 6-31G*/RHF geometry optimization with varied dielectric constant ε.

Parameter	A			B	C
	$\varepsilon = 1$	$\varepsilon = 10$	$\varepsilon = 78.3$	$\varepsilon = 1$	$\varepsilon = 1$
Bond Length, r, Å					
C^2—O^1	1.335	1.332	1.332	1.346	1.340
C^2—C^3	1.347	1.341	1.340	1.340	1.341
C^3—C^4	1.437	1.443	1.444	1.452	1.453
C^4—C^5	1.466	1.462	1.461	1.471	1.453
C^5—C^6	1.327	1.329	1.329	1.331	1.341
C^6—O^1	1.339	1.342	1.343	1.328	1.340
C^4—O^7	1.220	1.218	1.218	1.212	1.223
C^2—C^{19}	1.563	1.541	1.539	1.561	1.558
C^{19}—O^{20}	1.225	1.229	1.230	1.225	1.233
C^{19}—O^{21}	1.220	1.227	1.228	1.221	1.233
C^5—O^{12}	1.350	1.348	1.348	1.362	

(Table 2.2) contd.....

Parameter	A			B	C
	$\varepsilon = 1$	$\varepsilon = 10$	$\varepsilon = 78.3$	$\varepsilon = 1$	$\varepsilon = 1$
C^6—C^8					1.558
C^8—O^9					1.233
C^8—O^{10}					1.233
Bond Angle, ω, deg					
$O^1C^2C^3$	121.4	122.0	122.0	120.9	121.6
$C^2C^3C^4$	121.1	120.7	120.7	122.5	121.5
$C^3C^4C^5$	114.0	114.2	114.2	112.7	113.0
$C^4C^5C^6$	120.1	120.1	120.1	119.7	121.5
$C^5C^6O^1$	122.6	122.4	122.4	124.3	121.6
$C^6O^1C^2$	120.7	120.6	120.6	119.9	120.7
$C^4C^5O^{12}$	116.4	117.6	117.9	121.1	
$C^5C^4O^7$	120.4	119.9	120.2	122.3	123.5
$O^1C^2C^{19}$	114.6	114.0	114.0	114.4	115.8
$C^2C^{19}O^{21}$	114.5	115.6	115.8	114.5	115.8
$O^{20}C^{19}O^{21}$	132.7	129.9	129.4	132.5	130.6
$O^1C^6C^8$					115.8
$C^6C^8O^{10}$					115.8
$O^9C^8O^{10}$					130.6
Atomic Charge, q, a.u.					
O^1	-0.58	-0.61	-0.61	-0.59	-0.61
C^2	0.29	0.31	0.31	0.28	0.30
C^3	-0.39	-0.38	-0.38	-0.39	-0.40
C^4	0.56	0.56	0.56	0.57	0.59
C^5	0.23	0.24	0.24	0.21	-0.40
C^6	0.11	0.12	0.12	0.14	0.30
O^7	-0.70	-0.71	-0.71	-0.66	-0.72
C^8					0.74
O^9					-0.75
O^{10}					-0.71
O^{12}	-0.72	-0.79	-0.79	-0.66	
C^{19}	0.76	0.75	0.75	0.76	0.74
O^{20}	-0.72	-0.75	-0.76	-0.72	-0.75
O^{21}	-0.71	-0.76	-0.76	-0.71	-0.71

Ionization of the carboxyl groups substantially influences the geometry and electronic structure of the molecules in study. The charges on O^7, O^9, and O^{20} atoms sharply increase in the absolute value, while $q(C^3)$ and $q(C^5)$ become more negative. This redistribution of atomic charges primarily affects the geometry of the carboxyl groups. C^8—O^9 and C^{19}—O^{20} bonds are no longer conjugated with the system of C^2—C^3, C^4—O^7, and C^5—C^6 double bonds, which causes a strong elongation of C^2—C^{19} and C^6—C^8 bonds. The pyrone ring bond lengths also change: C^2—O^1 (A, B), C^2—C^3 (A-C), C^4—O^7 (A-C) bonds are elongated and C^3—C^4 (A-C), C^4—C^5 (C), C^6—O^1 (A, B) bonds are shortened. O^{12}—$H^{13}\cdots O^7$ hydrogen bond is retained in the anion of A. $r(O^7\cdots H^{13})$ distances are 2.075 ($\varepsilon = 1$), 2.171 ($\varepsilon = 10$), and 2.192 Å ($\varepsilon = 78.3$), which is slightly less than in the free acid A.

Sodium and Calcium Salts

Only molecules A-C are capable of forming salts with inorganic cations at physiological pH. Molecules A and B are monobasic acids; molecule C, dibasic. Along with electroneutral sodium salts, calcium salts of A and B with 1:1 acid–Ca^{2+} stoichiometry and that of C with 1:2 acid–Ca^{2+} stoichiometry were also examined to elucidate the effect of the nature of the salt-forming cation on molecular structure of gamma-pyrones. Calcium salts of A and B have the charge +1, while that of C, +2.

The geometric parameters and atomic charges for sodium and calcium salts of A-C, most differing from the corresponding values for the free acids, are presented in Tables **2.3** and **2.4**, respectively. The differences in steric and electronic structure in the series of sodium and calcium salts of A-C are mainly determined by the same factors as for the free acids. An increase of ε induces an increase of the distances between the cations and carboxyl oxygen atoms and a growth of positive charge on the metal atoms.

Table 2.3. Selected bond lengths, bond angles, and atomic charges in sodium salts of A-C obtained by full 6-31G*/RHF geometry optimization with varied dielectric constant ε.

Parameter	A			B	C
	$\varepsilon = 1$	$\varepsilon = 10$	$\varepsilon = 78.3$	$\varepsilon = 1$	$\varepsilon = 1$
Bond Length, *r*, Å					
C^2—O^1	1.330	1.330	1.330	1.338	1.339
C^2—C^3	1.336	1.336	1.336	1.331	1.331
C^3—C^4	1.452	1.448	1.448	1.465	1.466
C^4—C^5	1.467	1.461	1.461	1.473	1.466

(Table 2.3) contd.....

Parameter	A			B	C
	$\varepsilon = 1$	$\varepsilon = 10$	$\varepsilon = 78.3$	$\varepsilon = 1$	$\varepsilon = 1$
C^5—C^6	1.328	1.329	1.329	1.331	1.331
C^6—O^1	1.346	1.346	1.345	1.337	1.339
C^4—O^7	1.208	1.212	1.213	1.202	1.203
C^2—C^{19}	1.517	1.519	1.519	1.514	1.515
C^{19}—O^{20}	1.241	1.239	1.238	1.242	1.244
C^{19}—O^{21}	1.236	1.236	1.236	1.236	1.236
C^5—O^{12}	1.343	1.346	1.346	1.351	
C^6—C^8					1.515
C^8—O^9					1.244
C^8—O^{10}					1.236
O^9—Na^{11}					2.195
O^{10}—Na^{11}					2.195
O^{20}—Na^{22}	2.205	2.251	2.259	2.201	2.195
O^{21}—Na^{22}	2.204	2.264	2.271	2.204	2.195
Bond Angle, ω, deg					
$O^1C^2C^3$	123.2	122.9	122.9	122.8	123.3
$C^2C^3C^4$	120.1	120.1	120.1	121.3	120.6
$C^3C^4C^5$	113.9	114.4	114.4	112.8	113.0
$C^4C^5C^6$	120.3	120.1	120.2	119.8	120.6
$C^5C^6O^1$	122.7	122.4	122.4	124.3	123.3
$C^6O^1C^2$	119.8	120.2	120.2	119.0	119.2
$C^4C^5O^{12}$	117.0	117.3	117.4	120.9	
$C^5C^4O^7$	119.9	120.2	120.3	123.5	123.5
$O^1C^2C^{19}$	113.6	114.0	113.9	113.7	113.7
$C^2C^{19}O^{21}$	118.0	118.0	117.9	118.1	118.4
$O^{20}C^{19}O^{21}$	125.4	125.6	125.6	125.3	125.0
$O^1C^6C^8$					113.7
$C^6C^8O^{10}$					118.4
$O^9C^8O^{10}$					125.0
Atomic Charge, q, a.u.					
O^1	-0.60	-0.61	-0.61	-0.60	-0.62
C^2	0.31	0.33	0.33	0.30	0.30
C^3	-0.35	-0.35	-0.35	-0.35	-0.34

(Table 2.3) contd.....

Parameter	A			B	C
	$\varepsilon = 1$	$\varepsilon = 10$	$\varepsilon = 78.3$	$\varepsilon = 1$	$\varepsilon = 1$
C^4	0.54	0.56	0.56	0.56	0.58
C^5	0.25	0.25	0.25	0.24	-0.34
C^6	0.11	0.12	0.12	0.13	0.30
O^7	-0.64	-0.68	-0.69	-0.60	-0.61
C^8					0.84
O^9					-0.75
O^{10}					-0.72
Na^{11}					0.75
O^{12}	-0.77	-0.79	-0.79	-0.65	
C^{19}	0.84	0.85	0.85	0.84	0.84
O^{20}	-0.75	-0.75	-0.76	-0.75	-0.75
O^{21}	-0.73	-0.75	-0.75	-0.73	-0.72
Na^{22}	0.76	0.79	0.79	0.75	0.75

Table 2.4. Selected bond lengths, bond angles, and atomic charges in calcium salts of A-C obtained by full 6-31G*/RHF geometry optimization with varied dielectric constant ε.

Parameter	A			B	C
	$\varepsilon = 1$	$\varepsilon = 10$	$\varepsilon = 78.3$	$\varepsilon = 1$	$\varepsilon = 1$
Bond Length, r, Å					
C^2—O^1	1.327	1.328	1.328	1.334	1.345
C^2—C^3	1.332	1.333	1.333	1.328	1.328
C^3—C^4	1.465	1.456	1.455	1.476	1.476
C^4—C^5	1.468	1.462	1.461	1.474	1.476
C^5—C^6	1.330	1.329	1.330	1.332	1.328
C^6—O^1	1.353	1.346	1.345	1.347	1.345
C^4—O^7	1.200	1.210	1.211	1.195	1.190
C^2—C^{19}	1.490	1.500	1.501	1.487	1.493
C^{19}—O^{20}	1.257	1.248	1.247	1.258	1.255
C^{19}—O^{21}	1.251	1.245	1.245	1.252	1.249
C^5—O^{12}	1.335	1.343	1.345	1.343	
C^6—C^8					1.493
C^8—O^9					1.255
C^8—O^{10}					1.249

(Table 2.4) contd.....

Parameter	A			B	C
	$\varepsilon = 1$	$\varepsilon = 10$	$\varepsilon = 78.3$	$\varepsilon = 1$	$\varepsilon = 1$
O^9—Ca^{11}					2.266
O^{10}—Ca^{11}					2.293
O^{20}—Ca^{22}	2.264	2.325	2.334	2.262	2.266
O^{21}—Ca^{22}	2.264	2.325	2.334	2.261	2.293
Bond Angle, ω, deg					
$O^1C^2C^3$	124.4	123.8	123.7	119.9	124.4
$C^2C^3C^4$	119.4	119.6	119.6	124.3	120.2
$C^3C^4C^5$	113.9	114.1	114.2	118.4	112.9
$C^4C^5C^6$	120.4	120.4	120.4	120.7	120.2
$C^5C^6O^1$	122.7	122.5	122.4	124.6	124.4
$C^6O^1C^2$	119.2	119.7	119.7	113.3	117.9
$C^4C^5O^{12}$	117.5	117.4	117.6	120.3	
$C^5C^4O^7$	121.3	120.9	120.7	120.3	123.6
$O^1C^2C^{19}$	113.2	113.2	113.2		113.9
$C^2C^{19}O^{21}$	120.3	119.5	119.4		121.0
$O^{20}C^{19}O^{21}$	120.4	121.9	122.1		120.7
$O^1C^6C^8$				113.7	113.9
$C^6C^8O^{10}$				122.1	121.0
$O^9C^8O^{10}$				119.3	120.7
Atomic Charge, q, a.u.					
O^1	-0.61	-0.61	-0.61	-0.61	-0.64
C^2	0.31	0.33	0.33	0.30	0.29
C^3	-0.31	-0.32	-0.32	-0.31	-0.29
C^4	0.53	0.56	0.56	0.55	0.56
C^5	0.27	0.25	0.25	0.26	-0.29
C^6	0.10	0.12	0.13	0.12	0.29
O^7	-0.58	-0.66	-0.68	-0.55	-0.51
C^8					0.88
O^9					-0.79
O^{10}					-0.77
Ca^{11}					1.67
O^{12}	-0.75	-0.78	-0.78	-0.65	
C^{19}	0.88	0.89	0.89	0.88	0.88

(Table 2.4) contd.....

Parameter	A			B	C
	$\varepsilon = 1$	$\varepsilon = 10$	$\varepsilon = 78.3$	$\varepsilon = 1$	$\varepsilon = 1$
O^{20}	-0.79	-0.79	-0.79	-0.79	-0.79
O^{21}	-0.77	-0.78	-0.78	-0.77	-0.77
Ca^{22}	1.66	1.72	1.73	1.65	1.67

Comparison of the geometry and electronic parameters for isolated molecules of sodium and calcium salts of acids A-C shows that substitution of Ca^{2+} for Na^+ produces certain changes in molecular structure of the salts. These changes mostly relate to the structure of the carboxyl groups and are determined by a higher charge (1.65-1.67 a.u.) and electron-accepting ability of Ca^{2+}: C^8—O^9, C^8—O^{10}, C^{19}—O^{20}, C^{19}—O^{21} bonds become longer by 0.011-0.017 Å, C^2—C^{19} and C^6—C^8 bonds shorten by 0.021-0.029 Å, $O^9C^8O^{10}$ and $O^{20}C^{19}O^{21}$ angles decrease by ~ 4-5°, $C^2C^{19}O^{21}$ and $C^6C^8O^{10}$ angles increase by ~ 2-3°, and the absolute values of the charges on O^9, O^{10}, O^{20}, O^{21} atoms increase by 0.04-0.05 a.u. The structure of the pyrone ring slightly changes as well: the lengths of C^3—C^4, C^4—C^5, and C^6—O^1 bonds increase by less than 0.013 Å, C^4—O^7 and C^5—O^{12} bonds shorten by 0.007-0.019 Å, $q(C^3)$ and $q(C^5)$ become more positive by 0.02-0.03 a.u. The charge on O^7 atom sharply (by 0.05-0.11 a.u.) decreases in the absolute value. Hydrogen bonds in the molecules of salts slightly weaken: $r(O^7 \cdots H^{13})$ distances in O^{12}—$H^{13} \cdots O^7$ bonds in the salts of A are 2.147 (sodium salt) and 2.192 Å (calcium salt).

The effect of the cation nature on molecular structure of the salts is much less pronounced in the solvated state. Comparison of the geometry and electronic parameters of the molecules of sodium and calcium salts of A at $\varepsilon = 10$ and $\varepsilon = 78.3$ indicates that the differences relate exclusively to the structure of the carboxyl group, whereas the structure of the pyrone ring remains basically unchanged. The lengths of O^{12}—$H^{13} \cdots O^7$ hydrogen bonds are close to each other: $r(O^7 \cdots H^{13})$ distances are 2.172 and 2.180 Å (sodium salt, $\varepsilon = 10$ and $\varepsilon = 78.3$, respectively); 2.201 and 2.198 Å (calcium salt, $\varepsilon = 10$ and $\varepsilon = 78.3$, respectively).

The differences between the geometric parameters of the salts of A-C and the corresponding values for the free acids A-C mostly relate to the structure of the carboxyl groups, whereas the bond lengths and bong angles in the pyrone rings vary scarcely.

Complexes with Ca^{2+}

As it was mentioned above, the carbonyl and hydroxyl groups in adjacent positions of the pyrone ring can form chelate bonds with divalent cations. Such groups are present in molecules A and D. The ability of molecule B to form chelate complexes involving the methoxy and carbonyl groups was also studied. Selected bond lengths, bond angles, and atomic charges in the complexes of A, B, and D with Ca^{2+} are given in Table **2.5**. The data obtained indicate that substitution of the hydroxyl group in position 5 of the pyrone ring for the methoxy group in substance B does not hinder Ca^{2+} chelation. All heavy atoms in the calcium complex of B are located in the pyrone ring plane, in contrast to the free acid B, in the molecule of which the oxygen atom of the methoxy group is positioned out of this plane. Chelation of Ca^{2+} rules out the possibility for formation of O^{12}—$H^{13}\cdots O^7$ hydrogen bonds in the complex of A. For this reason, the geometric and electronic structures of isolated complexes vary rather slightly.

Table 2.5. Selected bond lengths, bond angles, and atomic charges in calcium chelate complexes of A, B, and D obtained by full 6-31G*/RHF geometry optimization with varied dielectric constant ε.

Parameter	A			B	D
	$\varepsilon = 1$	$\varepsilon = 10$	$\varepsilon = 78.3$	$\varepsilon = 1$	$\varepsilon = 1$
Bond Length, r, Å					
C^2—O^1	1.324	1.325	1.324	1.321	1.326
C^2—C^3	1.346	1.340	1.346	1.345	1.354
C^3—C^4	1.422	1.441	1.434	1.421	1.415
C^4—C^5	1.436	1.434	1.435	1.441	1.438
C^5—C^6	1.339	1.331	1.317	1.342	1.335
C^6—O^1	1.317	1.321	1.337	1.320	1.318
C^4—O^7	1.249	1.239	1.253	1.249	1.254
C^2—C^{19}	1.516	1.501	1.496	1.516	1.506
C^{19}—O^{20}	1.177	1.185	1.186	1.177	
C^{19}—O^{21}	1.303	1.309	1.319	1.303	
C^5—O^{12}	1.388	1.388	1.388	1.380	1.390
O^7—Ca^{17}	2.218	2.341	2.336	2.209	2.211
O^{12}—Ca^{17}	2.484	2.340	2.342	2.466	2.466
Bond Angle, ω, deg					
$O^1C^2C^3$	122.6	123.2	122.9	122.4	121.1
$C^2C^3C^4$	118.6	118.8	118.9	118.6	119.4
$C^3C^4C^5$	115.9	114.5	114.9	116.4	116.0
$C^4C^5C^6$	120.6	121.8	121.4	119.7	120.4
$C^5C^6O^1$	120.6	120.7	121.4	121.0	120.6
$C^6O^1C^2$	121.7	121.0	120.4	121.9	122.5

(Table 2.5) contd.....

Parameter	A			B	D
	$\varepsilon = 1$	$\varepsilon = 10$	$\varepsilon = 78.3$	$\varepsilon = 1$	$\varepsilon = 1$
$C^4C^5O^{12}$	114.0	113.6	113.0	114.4	114.3
$C^5C^4O^7$	119.0	119.4	121.0	119.3	118.7
$O^1C^2C^{19}$	115.6	113.8	114.7	115.8	112.6
$C^2C^{19}O^{21}$	112.1	112.2	113.1	112.2	
$O^{20}C^{19}O^{21}$	128.4	126.2	124.8	128.3	
$C^4O^7Ca^{17}$	126.1	120.6	119.1	124.9	125.6
$C^5O^{12}Ca^{17}$	113.6	118.6	117.9	113.2	113.5
$O^7Ca^{17}O^{12}$	67.2	67.4	68.4	68.2	67.8
Atomic Charge, *q*, a.u.					
O^1	-0.54	-0.57	-0.58	-0.54	-0.53
C^2	0.34	0.35	0.36	0.34	0.46
C^3	-0.30	-0.29	-0.29	-0.30	-0.36
C^4	0.65	0.65	0.66	0.63	0.65
C^5	0.22	0.25	0.26	0.26	0.21
C^6	0.22	0.21	0.22	0.21	0.22
O^7	-0.84	-0.80	-0.80	-0.84	-0.85
O^{12}	-0.89	-0.90	-0.88	-0.82	-0.90
Ca^{17}	1.77	1.80	1.80	1.76	1.76
C^{19}	0.81	0.82	0.82	0.81	-0.02
O^{20}	-0.48	-0.55	-0.56	-0.48	
O^{21}	-0.67	-0.69	-0.70	-0.67	

Solvation of calcium chelate of A results in several changes in the geometry and electronic structure of this molecule. Increase of negative charge on O^1 and O^{20} atoms invokes elongation of C^6—O^1, C^{19}—O^{20}, C^{19}—O^{21} bonds and shortening of C^2—C^{19}, C^5—C^6 bonds. Positive charge on Ca^{17} increases by 0.03 a.u., and this atom shifts toward O^{12}, which is reflected on $C^4O^7Ca^{17}$ and $C^5O^{12}Ca^{17}$ bond angles. The calcium atom slightly goes out of the pyrone ring plane: $C^4C^5O^{12}Ca^{17}$ torsion angles are 0.0°, 6.1°, and 8.2° at $\varepsilon = 1$, $\varepsilon = 10$, and $\varepsilon = 78.3$, respectively.

Examining the differences in the geometric parameters and atomic charges between the complexes and the free acids, the effect of complex formation on molecular structure can be traced. Introduction of electron-accepting calcium atom primarily results in the growth of negative charge on O^7 (by 0.13-0.29 a.u.) and O^{12} (by 0.10-0.20 a.u.) atoms. Positive charges on C^4 and C^6 atoms increase by ~0.1 a.u., $q(O^1)$ becomes less negative by ~0.05 a.u., and the charges on C^8 and C^{19} atoms increase by ~0.05 a.u. The most substantial changes in bond lengths, by 0.03-0.05 Å, occur in the vicinity of the calcium atom: C^4—O^7 and C^5—O^{12} bonds are weakened, C^3—C^4 and C^4—C^5 bonds are shortened. Displacement of π-electron density induced by formation of the chelate complex results in elongation of C^2—C^3 and C^5—C^6 bonds by ~0.01-0.02 Å, while C^6—O^1 bond length decreases by 0.025-0.035 Å. Weakening of π-conjugation also induces a certain elongation of C^2—C^{19}, C^2—C^{19} bonds and shortening of C^8—O^9, C^8—O^{10}, C^{19}—O^{20}, C^{19}—O^{21} bonds. Changes in bond lengths and atomic charges affect the bond angle values: $C^2O^1C^6$ and $C^3C^4C^5$ angles increase, $C^2C^3C^4$, $C^3C^2O^1$, $C^5C^6O^1$ angles decrease. As expected, the strongest changes are detected in $C^4C^5O^{12}$ and $C^5C^4O^7$ angles which define relative positions with respect to the pyrone ring of O^7 and O^{12} atoms directly involved in calcium chelation.

Sodium and Calcium Salts of Ca²⁺ Complexes

At physiological pH, only the calcium complexes of A and B can form salts with Na^+ and Ca^{2+}. The charge of the sodium salts of calcium complexes is +2; the charge of the calcium salts of the complexes, +3.

Data presented in Tables **2.6** and **2.7** indicate that the structures of isolated molecules of the salts of A and B calcium complexes are practically the same. Increase of ε mostly affects molecular structure of the sodium salt of calcium complex of A. Negative charges on O^{20}, O^{21} atoms of the carboxyl group and on O^1 atom of the pyrone ring increase, the absolute value of $q(O^7)$ decreases, and $q(Ca^{17})$ grows. These charge variations shift Na^{22} position closer toward O^{20} and O^{21} atoms by 0.015-0.030 Å, which is accompanied by elongation of C^{19}—O^{20}, C^{19}—O^{21} bonds by ~0.01 Å and shortening of C^2—C^{19}, C^5—O^{12} bonds. Decrease of the absolute value of negative charge on O^7 atom by 0.04 a.u. displaces Ca^{17} atom toward O^{12}, which is manifested in weakening of O^7—Ca^{17} bond and strengthening of O^{12}—Ca^{17} bond. Effect of the medium is weaker reflected on the geometry of the calcium salt of calcium complex of A. As in the sodium salt of the complex, Ca^{17} atom is displaced and negative charges on O^1, O^{20} and O^{21} atoms are increased by 0.02-0.04 a.u. Variation of ε has almost no effect on the structure of the pyrone rings in the salts of calcium complex of A, thus, geometry changes affect exclusively the structures of the carboxyl groups and position of the chelated calcium atom.

Variations in the geometry and electronic structure of the salts of calcium complexes of A and B upon substitution of Na^+ for Ca^{2+} are basically similar to those observed in the salts of the free acids A and B. The distances between chelated Ca^{17} cation and Ca^{22} or Na^{22} cations are very close to each other in the salts of calcium chelates of A and B: $r(Ca^{17}—Ca^{22}) = 9.54\text{-}9.58$ Å, $r(Ca^{17}—Na^{22}) = 9.36\text{-}9.44$ Å. It is suggested that these cations are responsible for interaction of the ligand with the conserved aspartates of opioid-like receptor binding pocket. It is important to note that the specified distances appear practically fixated in the series of molecules under consideration, which is required to provide complementarity of the ligand to active binding centers of the receptor.

Table 2.6. Selected bond lengths, bond angles, and atomic charges in sodium salts of calcium chelate complexes of A and B obtained by full 6-31G*/RHF geometry optimization with varied dielectric constant ε.

Parameter	A			B
	$\varepsilon = 1$	$\varepsilon = 10$	$\varepsilon = 78.3$	$\varepsilon = 1$
Bond Length, r, Å				
$C^2—O^1$	1.324	1.323	1.327	1.322
$C^2—C^3$	1.353	1.344	1.340	1.352
$C^3—C^4$	1.412	1.414	1.427	1.412
$C^4—C^5$	1.436	1.445	1.437	1.441
$C^5—C^6$	1.338	1.332	1.335	1.341
$C^6—O^1$	1.314	1.320	1.325	1.316
$C^4—O^7$	1.257	1.247	1.244	1.257
$C^2—C^{19}$	1.538	1.530	1.524	1.537
$C^{19}—O^{20}$	1.228	1.238	1.235	1.228
$C^{19}—O^{21}$	1.222	1.232	1.233	1.222
$C^5—O^{12}$	1.393	1.375	1.371	1.385
$O^7—Ca^{17}$	2.218	2.304	2.282	2.193
$O^{12}—Ca^{17}$	2.484	2.318	2.394	2.443
$Ca^{17}—Na^{22}$	9.414	9.438	9.429	9.389
$O^{20}—Na^{22}$	2.308	2.283	2.277	2.305
$O^{21}—Na^{22}$	2.293	2.278	2.279	2.291
Bond Angle, ω, deg				
$O^1C^2C^3$	121.7	123.1	122.3	121.5
$C^2C^3C^4$	119.2	119.0	119.3	119.2
$C^3C^4C^5$	115.8	115.3	115.5	116.4
$C^4C^5C^6$	120.6	121.2	120.5	119.6
$C^5C^6O^1$	120.6	120.4	121.1	121.0
$C^6O^1C^2$	122.1	121.0	121.3	122.3

(Table 2.6) contd.....

Parameter	A			B
	$\varepsilon = 1$	$\varepsilon = 10$	$\varepsilon = 78.3$	$\varepsilon = 1$
$C^4C^5O^{12}$	114.5	113.2	113.9	114.8
$C^5C^4O^7$	118.7	118.5	120.0	119.0
$O^1C^2C^{19}$	114.5	113.2	113.9	114.6
$C^2C^{19}O^{21}$	116.5	117.2	117.4	116.5
$O^{20}C^{19}O^{21}$	129.0	127.0	126.7	128.9
$C^4O^7Ca^{17}$	125.5	121.2	121.8	124.4
$C^5O^{12}Ca^{17}$	113.2	118.5	116.1	112.8
$O^7Ca^{17}O^{12}$	68.2	68.1	68.1	69.0
Atomic Charge, *q*, a.u.				
O^1	-0.52	-0.57	-0.57	-0.53
C^2	0.33	0.34	0.35	0.33
C^3	-0.32	-0.32	-0.32	-0.32
C^4	0.65	0.65	0.66	0.64
C^5	0.21	0.25	0.26	0.25
C^6	0.21	0.20	0.21	0.20
O^7	-0.86	-0.82	-0.82	-0.86
O^{12}	-0.90	-0.90	-0.89	-0.82
Ca^{17}	1.75	1.79	1.79	1.74
C^{19}	0.85	0.86	0.86	0.85
O^{20}	-0.70	-0.74	-0.74	-0.70
O^{21}	-0.68	-0.73	-0.74	-0.68
Na^{22}	0.82	0.80	0.80	0.81

Table 2.7. Selected bond lengths, bond angles, and atomic charges in calcium salts of calcium chelate complexes of A and B obtained by full 6-31G*/RHF geometry optimization with varied dielectric constant ε.

Parameter	A			B
	$\varepsilon = 1$	$\varepsilon = 10$	$\varepsilon = 78.3$	$\varepsilon = 1$
Bond Length, *r*, Å				
$C^2—O^1$	1.326	1.327	1.326	1.323
$C^2—C^3$	1.343	1.338	1.338	1.342
$C^3—C^4$	1.433	1.437	1.436	1.432
$C^4—C^5$	1.435	1.436	1.434	1.440

(Table 2.7) contd.....

Parameter	A			B
	$\varepsilon = 1$	$\varepsilon = 10$	$\varepsilon = 78.3$	$\varepsilon = 1$
C^5—C^6	1.340	1.334	1.334	1.343
C^6—O^1	1.319	1.322	1.324	1.321
C^4—O^7	1.244	1.239	1.240	1.244
C^2—C^{19}	1.516	1.505	1.505	1.515
C^{19}—O^{20}	1.242	1.244	1.242	1.242
C^{19}—O^{21}	1.236	1.239	1.242	1.237
C^5—O^{12}	1.383	1.372	1.368	1.375
O^7—Ca^{17}	2.246	2.282	2.309	2.235
O^{12}—Ca^{17}	2.509	2.428	2.382	2.489
Ca^{17}—Ca^{22}	9.572	9.533	9.551	9.540
O^{20}—Ca^{22}	2.354	2.346	2.346	2.350
O^{21}—Ca^{22}	2.341	2.342	2.343	2.339
Bond Angle, ω, deg				
$O^1C^2C^3$	122.7	122.8	122.9	122.5
$C^2C^3C^4$	118.6	119.1	119.1	118.6
$C^3C^4C^5$	115.7	115.2	115.2	116.3
$C^4C^5C^6$	120.6	120.5	120.7	119.5
$C^5C^6O^1$	120.8	121.6	121.4	121.3
$C^6O^1C^2$	121.6	120.9	120.8	121.8
$C^4C^5O^{12}$	113.8	114.4	114.1	114.3
$C^5C^4O^7$	119.3	120.5	119.9	119.6
$O^1C^2C^{19}$	113.4	113.4	113.1	113.5
$C^2C^{19}O^{21}$	119.1	119.1	118.9	119.2
$O^{20}C^{19}O^{21}$	123.1	123.0	123.0	123.0
$C^4O^7Ca^{17}$	126.4	121.9	121.0	125.1
$C^5O^{12}Ca^{17}$	114.3	114.9	116.6	113.7
$O^7Ca^{17}O^{12}$	66.3	68.0	67.8	67.3
Atomic Charge, q, a.u.				
O^1	-0.54	-0.57	-0.57	-0.55
C^2	0.33	0.34	0.35	0.33
C^3	-0.29	-0.29	-0.29	-0.29
C^4	0.65	0.65	0.65	0.63
C^5	0.23	0.26	0.26	0.27

(Table 2.7) contd.....

Parameter	A			B
	$\varepsilon = 1$	$\varepsilon = 10$	$\varepsilon = 78.3$	$\varepsilon = 1$
C^6	0.22	0.21	0.21	0.21
O^7	-0.82	-0.81	-0.81	-0.82
O^{12}	-0.88	-0.88	-0.88	-0.81
Ca^{17}	1.78	1.79	1.80	1.77
C^{19}	0.88	0.90	0.90	0.88
O^{20}	-0.75	-0.77	-0.77	-0.75
O^{21}	-0.73	-0.76	-0.77	-0.73
Ca^{22}	1.73	1.74	1.74	1.73

Energy Effects of Salt Formation and Ca^{2+} Chelation

Energy effects of formation of gamma-pyrone sodium and calcium salts and those of Ca^{2+} chelation were estimated according to the following scheme. The energy of salt formation was calculated as the energy of the salt molecule minus the energies of the acid anion and corresponding counterions. The energy of Ca^{2+} chelation was taken to be equal to the difference between the energy of the chelate molecule and the sum of the energies of the free acid molecule and Ca^{2+} cation. The energy of formation of the salt of the complex was estimated as the energy of this salt minus the energies of the acid anion, Ca^{2+}, and corresponding counterions.

The calculated energies of salt and complex formation are presented in Table **2.8**. The energy effects of Ca^{2+} chelation in the gas phase are close for all studied molecules. The energies of sodium and calcium salt formation in the gas phase practically coincide for the acids A and B. The energies of formation of salts of A and B calcium chelates are also close to each other. It is important that the energy of formation of the complex of B is comparable to those of A and D, according to our data. Therefore, when the hydroxyl group in position 5 of the pyrone ring is replaced with the methoxy group, chelation of Ca^{2+} does not become substantially less energetically favorable.

The energies of chelation and salt formation in the solvated phase ($\varepsilon = 10$ and $\varepsilon = 78.3$) were calculated for molecule A and its derivatives by two procedures. In the first case, we used the energies of molecules fully optimized in the framework of the PCM model, which regards the energy of a solvated molecule as the sum of the internal energy and the energy of electrostatic interaction of the molecule with the charges on the surface of the accommodating cavity. The second calculation procedure includes also the cavitation energy [37] and contributions of dispersion and exchange-repulsion forces [38] to the free energy of solvation. Supplementary geometry optimization of solvated molecules was not carried out in this case, so the PCM-optimized parameters were used. The latter procedure describes the energies of solvated molecules more correctly, but requires much more processing time. Table **2.8** demonstrates that introduction of additional corrections to the energies of solvated molecules results in a 2.4-3.3 kcal·mol^{-1} decrease in the absolute values of the energies of salt and complex formation and in a ~6 kcal·mol^{-1} decrease in the absolute energies of formation of chelate salts. As it should be expected, the absolute values of the energies of salt and complex formation strongly decrease upon solvation, but these processes remain energetically favorable.

Table 2.8. Energies of formation of sodium and calcium salts (ΔE_{salt}(Na), ΔE_{salt}(Ca)), calcium chelate complexes ($\Delta E_{complex}$(Ca)), and sodium and calcium salts of calcium chelate complexes ($\Delta E_{complex-salt}$ (Ca, Na), $\Delta E_{complex-salt}$ (Ca, Ca)) of A-D with varied dielectric constant ε.

Parameter	A			B	C	D
	$\varepsilon = 1$	$\varepsilon = 10$	$\varepsilon = 78.3$	$\varepsilon = 1$	$\varepsilon = 1$	$\varepsilon = 1$
ΔE_{salt}(Na), kcal·mol^{-1}	-138.0	-43.7 (-46.1)[a]	-34.2 (-36.9)	-139.2	-168.7	
ΔE_{salt}(Ca), kcal·mol^{-1}	-283.6	-89.0 (-92.0)	-70.6 (-73.7)	-286.0	-300.8	
$\Delta E_{complex}$(Ca), kcal·mol^{-1}	-105.5	-56.4 (-59.7)	-50.4 (-53.0)	-113.5		-114.9
$\Delta E_{complex-salt}$ (Ca, Na), kcal·mol^{-1}	-264.6	-105.9 (-112.0)	-91.2 (-97.2)	-273.8		
$\Delta E_{complex-salt}$ (Ca, Ca), kcal·mol^{-1}	-318.3	-142.6 (-148.6)	-126.6 (-132.7)	-328.7		

[a] Energies calculated without dispersion and exchange repulsion contributions are given in parentheses. Energies of formation of salts of C are presented on the basis of one salt bond.

Entropy changes accompanying the corresponding processes were not considered. However, taking into account the stability of calcium chelates of gamma-pyrone derivatives in aqueous solutions (log K = 2.5 for the calcium chelate of D [25] and similarity of the geometric and electronic structures of A-D and their derivatives illustrated by quantum-chemical calculations, it may be assumed that even if the entropy contributions to the energy of complex formation are of positive sign, their values are comparable in the series of studied molecules and do not substantially influence the absolute values of estimated energy effects. PCM calculations to take solvation effects into account were performed for comenic acid (substance A) and its derivatives only. Nevertheless, as the processes of salt formation and calcium chelation involving molecules B-D in the gas phase are more energetically favorable than the corresponding processes involving molecule A, it may be expected that the former processes should be allowed as well.

CONCLUSION

Quantum-chemical calculations demonstrate that molecules A-D in all examined forms share rather similar geometry and electronic structure. Several observed differences result from the nature of ring substituents, as well as from the presence or absence of intramolecular hydrogen bonds. Solvation of molecule A and its derivatives does not change their molecular structure noticeably, so it is safe to assume that solvation of molecules B-D would not fundamentally affect their geometry and electronic structure. However, our patch-clamp data indicate that molecules C and D, unlike molecules A and B, do not modulate the functioning of $Na_V1.8$ channels.

The structures of the carboxyl group in C and also of the carbonyl and hydroxyl groups in D do not differ substantially from the structures of the corresponding functional groups in A and B. The energies of formation of sodium and calcium salts of C and the energy of Ca^{2+} chelation by D are comparable with the energy effects of similar processes in A and B. Thus, in our opinion, inability of C and D to bind to the opioid-ilke receptor is determined by inability of these molecules to interact with one of inorganic cations due to the absence of required functional groups: the hydroxyl group in C (inability to chelate Ca^{2+}) and the carboxyl group in D (inability to form salts at physiological pH).

As opioid ligands usually contain positively charged functional groups that interact with the conserved aspartates of the receptors, it may be suggested that the molecules in study bind to the opioid-like receptor in the form of salts of calcium chelate complexes. Chelated Ca^{2+} cation forms an ion-ionic bond with one of the aspartates, while the other aspartate interacts with the cation that serves as a counterion for the carboxyl group of the ligand. The observed effect of B is in accordance with such mechanism of binding. Our calculations demonstrate that replacement of the hydroxyl group with the methoxy group in position 5 of the pyrone ring imposes no steric or energetic restrictions on Ca^{2+} chelation.

To provide the complementarity of the ligand to receptor binding pocket, the distances between bound cations in the ligand molecule should correspond to the distance between the conserved aspartates in the receptor. Intercationic distances in the salts of A and B calcium complexes were calculated to be equal to 9.4-9.6 Å. According to the model [22] that explains the role of inorganic cations in μ-opioid ligand binding, the distance between the aspartates in μ-opioid receptors is 9-11 Å, the intercationic distance in the complex of naloxone (a nonspecific opioid antagonist) with two Na^+ cations is 8.4 Å, the cation-cation distance in the complex of bremazocine (a nonselective opioid agonist) with two Na^+ cations is 8.9 Å. The intercationic distances in the salts of calcium chelates of A and B are comparable with the above distances.

It is necessary to point out that the results of quantum-chemical calculations demonstrate exclusively a principal possibility for the realization of the proposed ligand-receptor binding mechanism, namely, the absence of energetic and steric restrictions on formation of calcium chelate complexes and salts with inorganic cations. The stability of complexes in solution could not be evaluated. In the framework of our approach the molecules are considered to reside in a homogenous anisotropic milieu with the specified dielectric constant ε. This is just an approximation of the properties of receptor binding pocket, which should be ideally regarded as a heterogenous anisotropic medium with a fairly low value of ε, much more polarizable than the solvent (water) and, consequently, more capable of stabilizing the structures unstable in aqueous solutions. A more accurate approach to investigate the mechanisms of ligand-receptor binding of gamma-pyrone derivatives is direct docking of ligands to the binding pocket of the opioid-like receptor, but at present it does not seem possible, while the steric structure of the receptor remains unresolved. As ligand-receptor complexes of gamma-pyrones have not been isolated, it is also difficult to find a direct experimental evidence for the fact that the specified ligands interact with the receptor in the form of a salt of calcium chelate complex.

However, preclinical [39] and clinical [40] trials of comenic acid (molecule A) have revealed a remarkable feature of this substance: it has no addictive properties and other adverse side effects being applied both on humans and animals. Therefore, it may be assumed that comenic acid does not interact with μ-opioid receptors and is absolutely safe for administration in humans. The molecule of comenic acid is practically planar, it has a very simple structure and contains only a few functional groups which could be involved in ligand-receptor binding. The structure is also rather rigid, so the positions of these functional groups can hardly be adjusted. We hypothesize that affinity of comenic acid to μ-opioid receptor is too low to activate it. In other words, the energy of ligand-receptor binding of comenic acid to μ-opioid receptor is insufficient to trigger the conformational change in the receptor molecule. On the other hand, both comenic acid and morphine can activate the opioid-like receptor coupled to $Na_v1.8$ channels, but affinity of morphine to this receptor is higher than that of comenic acid ($K_d = 8$ and 100 nM, respectively) [14, 18], and the morphine effect is virtually irreversible [18]. Most probably, opioid-like receptor binding pocket is rather small and has a slit-like shape, so the planar molecule of comenic acid is accommodated rather snugly. The morphine molecule occupies the whole volume of the binding pocket, forming a very stable ligand-receptor complex. It is also very important to note that endogenous μ-opioid agonists endomorphin-1 and endomorphin-2 did not exhibit any $Na_v1.8$ channel-modulating activity [41], which indicates that these molecules are too large to be accommodated in opioid-like receptor binding pocket.

The morphine molecule contains a nitrogen atom directly involved in binding to μ-opioid receptors. To find out how the change of the heteroatom in the pyrone ring could influence the ability of gamma-pyrones to activate the opioid-like receptor, several gamma-pyridone derivatives which have a nitrogen atom in the heterocyclic ring instead of an oxygen atom in gamma-pyrones were synthesized and investigated by both patch-clamp and quantum-chemical methods.

CONFLICT OF INTEREST

The authors confirm that they have no conflict of interest to declare for this publication.

ACKNOWLEDGEMENTS

Declared none.

REFERENCES

[1]   Emami S, Hosseinimehr SJ, Taghdisi SM, Akhlaghpoor S. Kojic acid and its manganese and zinc complexes as potential radioprotective agents. Bioorg Med Chem Lett 2007; 17(1): 45-8.
[http://dx.doi.org/10.1016/j.bmcl.2006.09.097] [PMID: 17049858]

[2]   Hosseinimehr SJ, Emami S, Zakaryaee V, Ahmadi A, Moslemi D. Radioprotective effects of kojic acid against mortality induced by gamma irradiation in mice. Saudi Med J 2009; 30(4): 490-3.
[PMID: 19370273]

[3]   Wang K, Liu C, Di CJ, *et al.* Kojic acid protects C57BL/6 mice from gamma-irradiation induced damage. Asian Pac J Cancer Prev 2014; 15(1): 291-7.
[http://dx.doi.org/10.7314/APJCP.2014.15.1.291] [PMID: 24528043]

[4]   Tanaka R, Tsujii H, Yamada T, *et al.* Novel 3α-methoxyserrat-14-en-21β-ol (PJ-1) and 3β-methoxyserrat-14-en-21β-ol (PJ-2)-curcumin, kojic acid, quercetin, and baicalein conjugates as HIV agents. Bioorg Med Chem 2009; 17(14): 5238-46.
[http://dx.doi.org/10.1016/j.bmc.2009.05.049] [PMID: 19515569]

[5]   Wei Y, Zhang C, Zhao P, Yang X, Wang K. A new salicylic acid-derivatized kojic acid vanadyl complex: synthesis, characterization and anti-diabetic therapeutic potential. J Inorg Biochem 2011; 105(8): 1081-5.
[http://dx.doi.org/10.1016/j.jinorgbio.2011.05.008] [PMID: 21726771]

[6]   Aytemir MD, Septioğlu E, Caliş U. Synthesis and anticonvulsant activity of new kojic acid derivatives. Arzneimittelforschung 2010; 60(1): 22-9.
[PMID: 20184223]

[7]   Chisvert A, Sisternes J, Balaguer A, Salvador A. A gas chromatography-mass spectrometric method to determine skin-whitening agents in cosmetic products. Talanta 2010; 81(1-2): 530-6.
[http://dx.doi.org/10.1016/j.talanta.2009.12.037] [PMID: 20188958]

[8]   Kwak SY, Choi HR, Park KC, Lee YS. Kojic acid-amino acid amide metal complexes and their melanogenesis inhibitory activities. J Pept Sci 2011; 17(12): 791-7.
[http://dx.doi.org/10.1002/psc.1404] [PMID: 21957050]

[9]   Liou SS, Shieh WL, Cheng TH, Won SJ, Lin CN. Gamma-pyrone compounds as potential anti-cancer drugs. J Pharm Pharmacol 1993; 45(9): 791-4.
[http://dx.doi.org/10.1111/j.2042-7158.1993.tb05686.x] [PMID: 7903365]

[10] Kayser O, Kiderlen AF, Croft SL. Antileishmanial activity of two gamma-pyrones from *Podolepis hieracioides* (*Asteraceae*). Acta Trop 2003; 86(1): 105-7.
[http://dx.doi.org/10.1016/S0001-706X(02)00258-9] [PMID: 12711109]

[11] Sander K, Kottke T, Weizel L, Stark H. Kojic acid derivatives as histamine H_3 receptor ligands. Chem Pharm Bull (Tokyo) 2010; 58(10): 1353-61.
[http://dx.doi.org/10.1248/cpb.58.1353] [PMID: 20930404]

[12] Chumasov EI, Shurygin AYa , Soldatova SYu , Svetikova KM, Drobyshevskii AI. Regeneration of nerves under the influence of balyse-2 and lactovit. Neurosci Behav Physiol 1994; 24(6): 452-6.
[http://dx.doi.org/10.1007/BF02360164] [PMID: 7715762]

[13] Demina NI, Zlishcheva LI, Shurygin AYa. The effect of Baliz-2 on the regeneration of rabbit ear skin after radiation-induced and mechanical injury. Dokl Biol Sci 2000; 372: 293-5.
[PMID: 10944727]

[14] Derbenev AV, Krylov BV, Shurygin AYa. Effects of meconic and comenic acids on slow sodium channels of secondary neurons. Membr Cell Biol 2000; 13(3): 379-87.
[PMID: 10768488]

[15] Acton QA, Ed. Hyperalgesia: new insights for the healthcare professional. 2013 ed., Atlanta: ScholarlyEditions 2013.

[16] Hamill OP, Marty A, Neher E, Sakmann B, Sigworth FJ. Improved patch-clamp techniques for high-resolution current recording from cells and cell-free membrane patches. Pflügers Arch 1981; 391(2): 85-100.
[http://dx.doi.org/10.1007/BF00656997] [PMID: 6270629]

[17] Sakmann B, Neher E, Eds. Single-channel recording. 2nd ed., New York, Dordrecht, Heidelberg, London: Springer 2009.

[18] Krylov BV, Derbenev AV, Podzorova SA, Liudyno MI, Kuzmin AV, Izvarina NL. [Morphine decreases the voltage sensitivity of the slow sodium channels]. Ross Fiziol Zh Im I M Sechenova 1999; 85(2): 225-36.
[PMID: 10389179]

[19] Strader CD, Fong TM, Tota MR, Underwood D, Dixon RA. Structure and function of G protein-coupled receptors. Annu Rev Biochem 1994; 63(1): 101-32.
[http://dx.doi.org/10.1146/annurev.bi.63.070194.000533] [PMID: 7979235]

[20] Baldwin JM, Schertler GF, Unger VM. An alpha-carbon template for the transmembrane helices in the rhodopsin family of G-protein-coupled receptors. J Mol Biol 1997; 272(1): 144-64.
[http://dx.doi.org/10.1006/jmbi.1997.1240] [PMID: 9299344]

[21] Pogozheva ID, Lomize AL, Mosberg HI. Opioid receptor three-dimensional structures from distance geometry calculations with hydrogen bonding constraints. Biophys J 1998; 75(2): 612-34.
[http://dx.doi.org/10.1016/S0006-3495(98)77552-6] [PMID: 9675164]

[22] Zhorov BS, Ananthanarayanan VS. Homology models of μ-opioid receptor with organic and inorganic cations at conserved aspartates in the second and third transmembrane domains. Arch Biochem Biophys 2000; 375(1): 31-49.
[http://dx.doi.org/10.1006/abbi.1999.1529] [PMID: 10683246]

[23] Bryant BE, Fernelius WC. Some metal complexes of kojic acid. J Am Chem Soc 1954; 76(21): 5351-2.
[http://dx.doi.org/10.1021/ja01650a026]

[24] Petrola R. Spectrophotometric study on the equilibrium of substituted 3-hydroxy-4H-pyran-4-ones with Zn(II) ions in aqueous solution. Finn Chem Lett 1985; 12(5): 219-24.

[25] Okač A, Kolařik Z. Potentiometrische Untersuchung von Komplexsalzen der Kojisäure in wässrigen Lösungen. Collect Czech Chem Commun 1959; 24(1): 266-72.
[http://dx.doi.org/10.1135/cccc19590266]

[26] Zhorov BS, Ananthanarayanan VS. Similarity of Ca(2+)-bound conformations of morphine and Met-enkephalin: a computational study. FEBS Lett 1994; 354(2): 131-4.
[http://dx.doi.org/10.1016/0014-5793(94)01071-4] [PMID: 7957911]

[27] Zhorov BS, Ananthanarayanan VS. Conformational analysis of the Ca(2+)-bound opioid peptides: implications for ligand-receptor interaction. J Biomol Struct Dyn 1995; 13(1): 1-13.
[http://dx.doi.org/10.1080/07391102.1995.10508817] [PMID: 8527022]

[28] Zhorov BS, Ananthanarayanan VS. Conformational and electrostatic similarity between polyprotonated and Ca(2+)-bound μ-opioid peptides. J Biomol Struct Dyn 1996; 14(2): 173-83.
[http://dx.doi.org/10.1080/07391102.1996.10508106] [PMID: 8913853]

[29] Andrianov GN, Puyal J, Raymond J, Ventéo S, Demêmes D, Ryzhova IV. Immunocytochemical and pharmacological characterization of metabotropic glutamate receptors of the vestibular end organs in the frog. Hear Res 2005; 204(1-2): 200-9.
[http://dx.doi.org/10.1016/j.heares.2005.02.003] [PMID: 15925205]

[30] Hariharan P, Pople J. The influence of polarization functions on molecular orbital hydrogenation energies. Theor Chim Acta 1973; 28(3): 213-22.
[http://dx.doi.org/10.1007/BF00533485]

[31] Schmidt M, Baldridge K, Boatz J, *et al.* General atomic and molecular electronic structure system. J Comput Chem 1993; 14(11): 1347-63.
[http://dx.doi.org/10.1002/jcc.540141112]

[32] Tomasi J, Persico M. Molecular interactions in solution: an overview of methods based on continuous distributions of the solvent. Chem Rev 1994; 94(7): 2027-94.
[http://dx.doi.org/10.1021/cr00031a013]

[33] Petrola R. Spectrophotometric study on the equilibria of pyromeconic acid derivatives with proton in aqueous solution. Finn Chem Lett 1985; 12(5): 207-12.

[34] Petrola R. Stability of yttrium (iii) complexes of substituted 3-hydroxy-4H-pyran-4-ones in aqueous solution. Finn Chem Lett 1986; 13(5): 129-35.

[35] Miyamoto S, Brochmann-Hanssen E. Dissociation constants of certain γ-pyrone dicarboxylic acids. Meconic acid and chelidonic acid. J Pharm Sci 1962; 51(6): 552-4.
[http://dx.doi.org/10.1002/jps.2600510613] [PMID: 14474811]

[36] Petrola R. UV absorption-spectra and acid strengths of kojic acid and its ring-substituted derivatives in aqueous solution. Finn Chem Lett 1985; 12(5): 201-6.

[37] Langlet J, Claverie P, Caillet J, Pullman A. Improvements of the continuum model. 1. Application to the calculation of the vaporization thermodynamic quantities of nonassociated liquids. J Phys Chem 1988; 92(6): 1617-31.
[http://dx.doi.org/10.1021/j100317a048]

[38] Amovilli C, Mennucci B. Self-consistent-field calculation of Pauli repulsion and dispersion contributions to the solvation free energy in the polarizable continuum model. J Phys Chem B 1997; 101(6): 1051-7.
[http://dx.doi.org/10.1021/jp9621991]

[39] Plakhova V, Rogachevsky I, Lopatina E, *et al.* A novel mechanism of modulation of slow sodium channels: from ligand-receptor interaction to design of an analgesic medicine. Act Nerv Super Rediviva 2014; 56(3-4): 55-64.

[40] Lopatina EV, Polyakov YuI. Synthetic analgesic Anoceptin: results of preclinical and clinical trials. Efferent Therapy 2011; 17(3): 79-81.

[41] Katina IE, Shchegolev BF, Zadina JE, McKee ML, Krylov BV. Endomorphins inhibit currents through voltage-dependent sodium channels. Sensornye Sistemi 2003; 17(1): 7-23.

CHAPTER 3

Possible Mechanisms of Binding of Gamma-Pyridones to the Opioid-Like Receptor

Abstract: The nociceptive system codes noxious signals by increasing its impulse firing. $Na_V1.8$ channels play a central role in the process of primary sensory coding. The Almers' method is almost ideal for the study of behavior of their gating device. Application of this method makes it possible to elucidate the mechanism of receptor-coupled modulation of $Na_V1.8$ channels by opioid-like receptors, which exhibit high affinity to some gamma-pyrone and gamma-pyridone derivatives. A remarkable feature characterizing these substances is their ability to chelate calcium. That is why opioid-like receptors recognize these attacking molecules in physiologically appropriate conditions by activation of a very important additional mechanism of ion-ionic interactions switched on by attacking molecules with chelated calcium. This conclusion is confirmed by the study of the effects of gamma-pyridone derivatives, which are structurally very close to gamma-pyrones and, in addition, also have an ability to chelate calcium.

The results discussed in this and the previous chapters open a new approach to solve the problem of recognition of medicinal analgesic substances. Our quantum-chemical calculations demonstrate that calcium chelation process plays an important role in ligand-receptor binding and it is energetically allowed not only in vacuum but also in the adequate physiological environment. Conclusions concerning the probable structure of opioid-like receptor binding pocket are presented.

Keywords: Ca^{2+} chelate complex, Gamma-pyridone derivatives, Limiting slope procedure, $Na_V1.8$ channels, Nociception, Opioid-like receptor, Patch-clamp method, Quantum-chemical calculations.

The main result obtained in Chapter **2** of this book is elucidation of the physiological role of calcium chelation by gamma-pyrone derivatives, which is fundamentally important for ligand-receptor binding. Participation of the calcium ion in binding of gamma-pyrones to the opioid-like receptor is completely not obvious, and it allows introducing a new approach to analyze physiological activity of potential analgesics. This conclusion can be additionally verified by investigation of physiological effects of molecules relating to a class of substances, similar in structure to gamma-pyrones. We have chosen for our

Boris V. Krylov, Ilia V. Rogachevskii, Tatiana N. Shelykh, Vera B. Plakhova

further studies gamma-pyridone derivatives, which are distinguished from gamma-pyrones by the nature of the ring heteroatom (gamma-pyridones contain a nitrogen atom, as opposed to an oxygen atom in gamma-pyrones).

Many gamma-pyridones are demonstrated to be of pharmacological importance: they can exhibit analgesic activity [1 - 5], display anti-inflammatory [3, 4, 6], antitumor [7, 8], antibacterial [9, 10], antimicrobial [11], antimalarial effects [12], positively influence the cardiovascular system [13] and can also be applied for the treatment of Parkinson's disease [14].

3-Hydroxy-gamma-pyridone derivatives are widely used in various fields of medicinal chemistry to treat the diseases caused by excess microelements in serum [15 - 17], as well as to design radioactive and fluorescent labels for diagnostics [17]. These applications of 3-hydroxy-gamma-pyridones are based on their ability to chelate doubly and triply charged cations (Al^{3+}, Fe^{3+}, Ga^{3+}, Zn^{2+}, Cu^{2+}, Ca^{2+}, Mg^{2+}) [17 - 20] through the carbonyl and hydroxyl groups at contiguous positions of the pyridone ring.

PATCH-CLAMP INVESTIGATION OF GAMMA-PYRIDONES

Insofar as gamma-pyridones are structurally related to gamma-pyrones discussed in Chapter **2**, it is of interest to investigate the effect of the ring heteroatom on physiological and structural properties of these compounds in order to find a new mechanism or new agents capable of acting as analgesics by selective modulation of $Na_V1.8$ channels. Two molecules were chosen to examine their ability to decrease the effective charge of $Na_V1.8$ channel activation gating device: 5-hydroxy-1-methyl-gamma-pyridone-2-carboxylic acid (substance E) and 5-hydroxy-2-hydroxymethyl-gamma-pyridone (substance F). Their structural formulae are presented in Fig. (**3.1**). It is important that the latter compound is a structural analog of kojic acid (substance D, Fig. **2.8**) which showed no appreciable activity in our patch-clamp experiments, while the former one is a structural analog of comenic acid (substance A, Fig. **2.8**) which is capable to activate the opioid-like receptor being applied at nanomolar concentrations.

The families of $Na_V1.8$ currents in the control experiment and after extracellular application of substance E are presented in Fig. (**3.2.a**). It is clearly seen that the amplitude values of the currents are reduced, indicating that the decrease of the channels density may take place. The peak current-voltage curve shifts in the depolarizing direction after substance E has been applied (Fig. **3.2.b**). The voltage dependencies of normalized $G_{Na\_s}(E)$ functions differ between the control and substance E data at negative potentials. In the latter case this function is steeper (Fig. **3.3.a**). When $G_{Na\_s}(E)$ dependencies are obtained, the Almers' limiting slope procedure can be applied, making it possible to evaluate Z_{eff} at the most negative potentials (Fig. **3.3.b**). Our experimental results demonstrate that substance F is also able to decrease Z_{eff} of Na$_V$1.8 channel activation gating system. Its action is characterized by only two manifestations of inhibitory properties: it reduces the amplitude values of the currents (Fig. **3.4.a**) and decreases Z_{eff} (Fig. **3.5**). Somewhat unexpected in this case is the lack of the voltage shift of Na$_V$1.8 channels activation gating process (Fig. **3.4.b**). On the contrary, the effects of substance E are absolutely of the same character as those observed after morphine or comenic acid application (Figs. **1.13**, **2.2**). The decrease in Z_{eff} after extracellular application of substance E occurs due to activation of the receptor-coupled membrane mechanism (Fig. **1.17**). Indeed, this conclusion is based on the fact that nonspecific opioid antagonist naloxone (NLX) switched off the effect of substance E (Fig. **3.6**). Fig. (**3.7**) summarizes the effects of investigated gamma-pyridone derivatives. Both substances E and F decrease Z_{eff}, which makes it possible to predict their antinociceptive action on the organismal level.

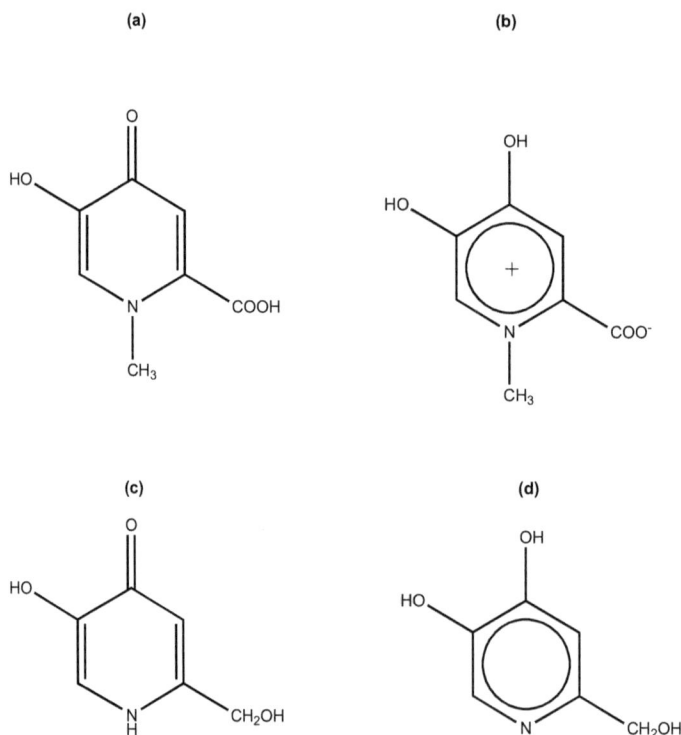

Fig. (3.1). Structural formulae of gamma-pyridones.
a – 5-hydroxy-1-methyl-gamma-pyridone-2-carboxylic acid (substance E), NH-form;
b – 5-hydroxy-1-methyl-gamma-pyridone-2-carboxylic acid (substance E), OH-form;
c – 5-hydroxy-2-hydroxymethyl-gamma-pyridone (substance F), NH-form;
d – 5-hydroxy-2-hydroxymethyl-gamma-pyridone (substance F), OH-form.

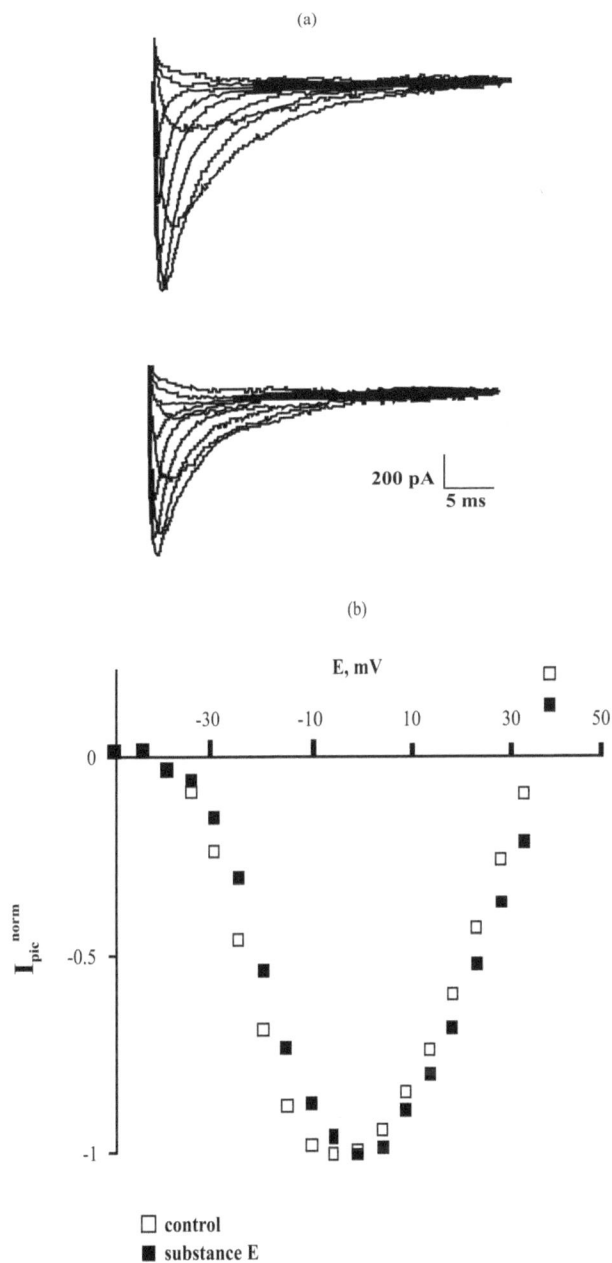

Fig. (3.2). Effects of substance E on $Na_V1.8$ channels.
a – Families of sodium currents measured in the control experiment (top) and after application of substance E at 1 μM (bottom);
b – Positive shift of the normalized peak current-voltage curve after application of substance E.

(a)

(b)

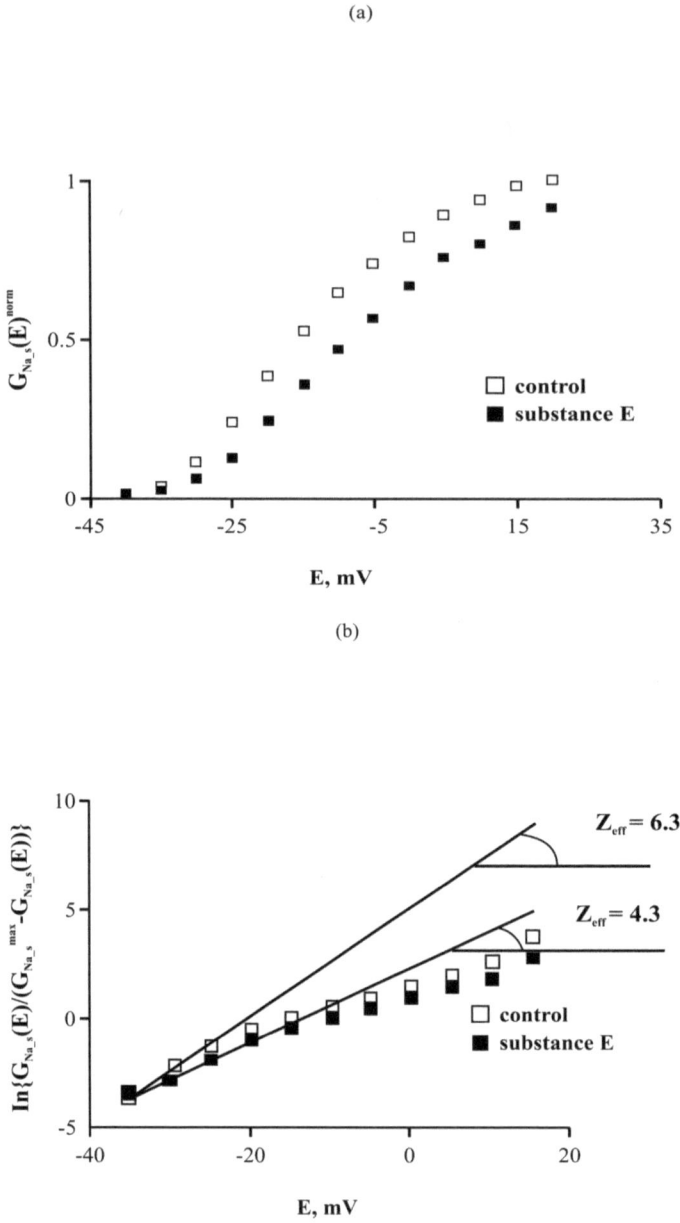

Fig. (3.3). Decrease of effective charge of $Na_V1.8$ channels activation gating device after application of substance E.

a – Voltage dependence of the normalized peak conductance used for Z_{eff} evaluation;

b – Z_{eff} evaluation by the Almers' limiting slope procedure in the control experiment and after application of substance E at 1 μM .

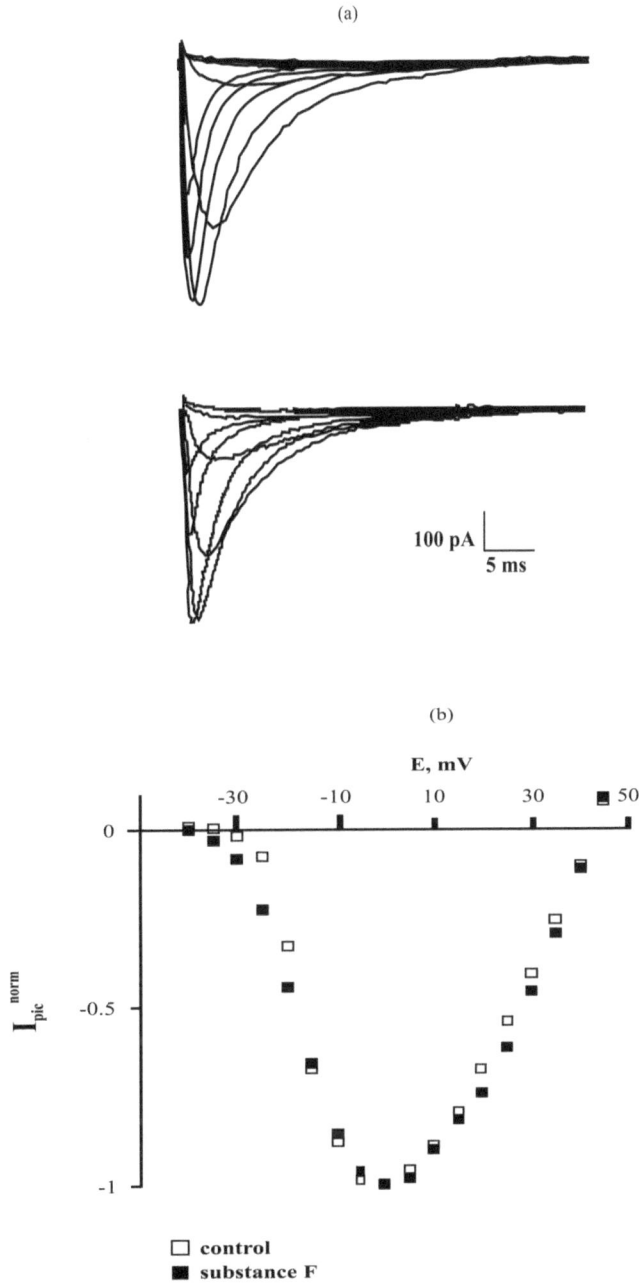

Fig. (3.4). Effects of substance F on $Na_V1.8$ channels.
a – Families of sodium currents measured in the control experiment (top) and after application of substance F at 1 μM (bottom);
b – The normalized peak current-voltage curve after application of substance F.

(a)

(b)

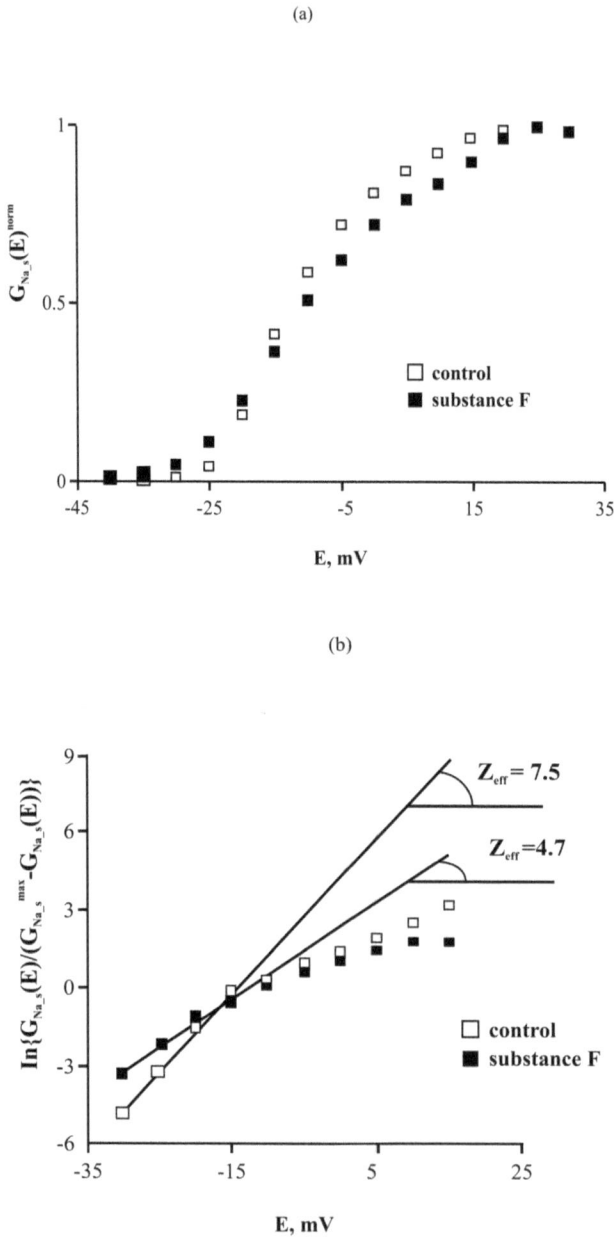

Fig. (3.5). Decrease of effective charge of $Na_V1.8$ channels activation gating device after application of substance F.

a – Voltage dependence of the normalized peak conductance used for Z_{eff} evaluation;
b – Z_{eff} evaluation by the Almers' limiting slope procedure in the control experiment and after application of substance F at 1 μM.

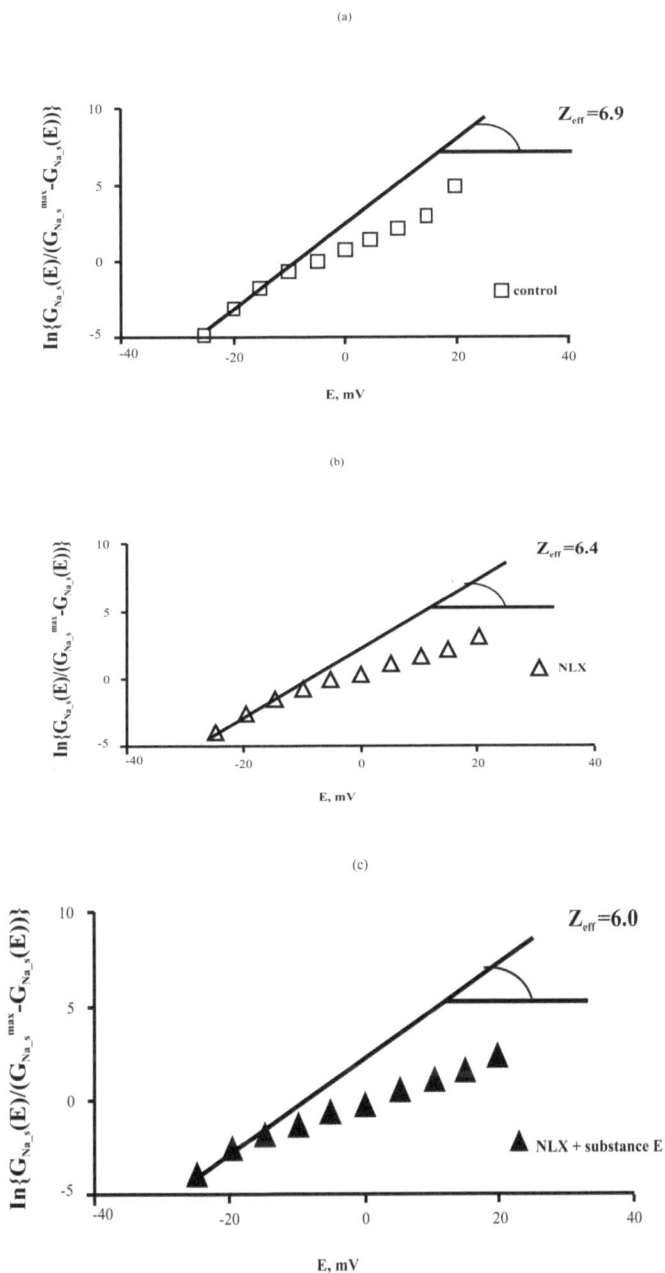

Fig. (3.6). Substance E activates the opioid-like receptor mechanism.
a – The control value of Z_{eff} (white squares);
b – Z_{eff} after NLX (50 μM) application (white triangles);
c – Z_{eff} after combined application of NLX and substance E (black triangles).

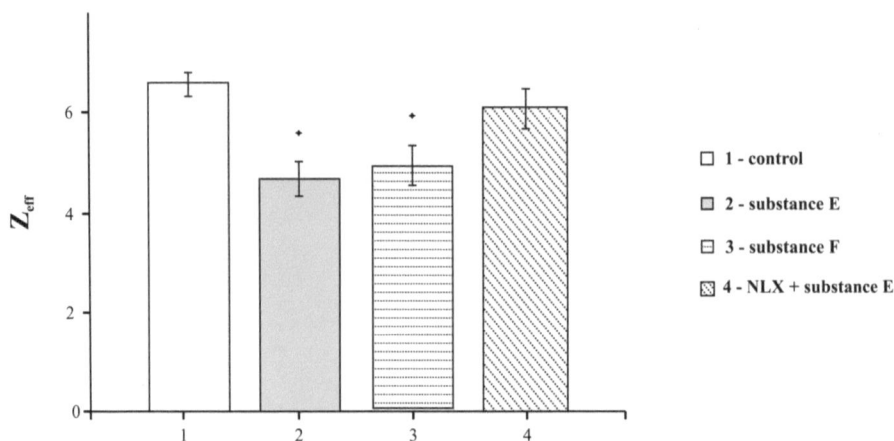

Fig. (3.7). Effects of gamma-pyridone derivatives on $Na_V1.8$ channels.
1 – The control value of Z_{eff} (6.5± 0.2, n = 14);
2 – Z_{eff} is reduced to 4.7±0.3 (n = 14) after application of substance E at 1 μM;
3 – Application of substance F decreases Z_{eff} to 4.9±0.4 (n = 16);
4 – Combined application of NLX (50 μM) and substance E (1 μM) does not result in a statistically significant decrease of Z_{eff} (6.1±0.4, n = 18).
*– difference between experimental and control data is statistically significant (in the cases of substances E and F application).

The observed effect of both gamma-pyridone and gamma-pyrone derivatives is blocked by nonspecific opioid antagonists, indicating the existence of the same receptor site responsible for binding the agents and common mechanism of interaction of these compounds with the opioid-like receptor. General structural similarity between gamma-pyridones and gamma-pyrones also suggests that these molecules should exhibit analogous physiological properties. However, the ability of substance F to modulate $Na_V1.8$ channel activation gating system and inability of substance D to do so indicates that replacement of the oxygen atom in the heteroring by the nitrogen atom may influence the physiological effect in some cases.

QUANTUM-CHEMICAL STUDY OF GAMMA-PYRIDONES

In order to support the experimental patch-clamp data we performed a quantum-chemical study on steric and electronic structure of substances D-F in various molecular forms, estimated their ability to interact with inorganic cations, examined the effect of the heteroatom on structure of the studied molecules by comparing their calculated parameters, and proposed a feasible mechanism for ligand-receptor binding on the basis of the obtained data. Molecule A was examined in detail in Chapter **2**.

The geometric parameters of neutral molecules D-F (free acids), their anions, calcium salts, calcium chelates, and calcium salts of calcium chelates were fully optimized *ab initio* by RHF method with 6-31G* basis set [21] using the GAMESS program [22]. The calculations were performed both for isolated molecules (dielectric constant $\varepsilon = 1$) and with account taken of solvation in terms of the polarized continuum model (PCM) [23] with $\varepsilon = 78.3$ (simulation of aqueous solution) and $\varepsilon = 10$ (simulation of receptor binding pocket milieu).

Depending on pH, molecules A, D, E, and F can exist in different molecular forms: neutral, conjugate acid (*i.e.*, protonated at the nitrogen or carbonyl oxygen atom), and conjugate base (the hydroxy group in position 5 of the heterocycle is ionized). According to published data, the acidity constants pK_1 of conjugate acids and pK_2 of neutral forms are as follows: $pK_2 = 7.8$ (A) [24], $pK_2 = 7.9$ (D) [25], $pK_2 = 9.0$ (E) [26], and $pK_1 = 3.3$ (F) [27]. There are no data available on pK_1 of substances A and D, but we assume that these values are of the same order as pK_1 of 3-hydroxy-2-methyl-gamma-pyrone ($pK_1 = 2.3$) [28]. Taking into account that our patch-clamp experiments were carried out at physiological pH (7.4), the cationic forms of A, D, E, and F should not be considered, as they are stable only in strongly acidic media. As we hope to demonstrate that the process of calcium chelation by molecules in study is energetically favorable, the conjugated bases were not investigated as well, though their contribution to the equilibrium between possible molecular forms should not be neglected at physiological pH. Obviously, negative charge on the oxygen atom of deprotonated hydroxyl group favors calcium chelation, which means that studying this process with participation of neutral molecules should make it possible to estimate the minimal absolute enthalpy of the process.

5-Hydroxy-2-hydroxymethyl-gamma-pyridone (substance F) can exist as two tautomers: 4-oxo (NH-form, Fig. **3.1.c**) and 4-hydroxy (OH-form, Fig. **3.1.d**). Analogous equilibria involving gamma-pyridone and its various derivatives have been studied in detail [27, 29 - 32]. Neutral molecule F was found to exist mainly as 4-oxo tautomer in aqueous solution [27], and the equilibrium constant between NH- and OH-tautomers of homologous 2-hydroxymethyl-5-methoxy-gamma-pyridone was estimated (log $K_t = 2.8$) [31]. We performed full *ab initio* RHF/6-31G* geometry optimization of both NH- and OH-tautomers of the free acid F and its 1:1 calcium chelate. NH-tautomers were calculated to be more stable than the corresponding OH-forms, the differences in the total energies being 1.6, 5.7 and 6.5 kcal·mol^{-1} for the free acid and 44.2, 27.5 and 25.3 kcal·mol^{-1} for the calcium chelate complex at $\varepsilon = 1$, 10 and 78.3, respectively.

As a rule, N-substituted gamma-pyridones are incapable of undergoing tautomeric transformations because of the lack of a labile proton, but this role is played by the hydrogen atom of the carboxyl group in molecule E. X-ray diffraction data for carboxypyridone E [33] and some other potentially tautomeric N-substituted hydroxypyridones [33, 34] demonstrate that all compounds studied exist as OH-tautomers (Fig. **3.1.b**) in crystal lattice, which indicates that OH-form is likely more energetically favorable in the solid state. As follows from the total energies of NH- and OH-tautomers of molecule E optimized at RHF/6-31G* level, NH-tautomer (Fig. **3.1.a**) is more stable than OH-tautomer both in the gas phase and in the solvated state by 23.9, 8.1 and 5.4 kcal·mol^{-1} at $\varepsilon = 1$, 10 and 78.3, respectively. The energy differences between NH- and OH-forms of 1:1 calcium chelate of molecule F were 88.7, 37.7 and 30.4 kcal·mol^{-1} at $\varepsilon = 1$, 10 and 78.3, respectively. A noticeable difference in the energies of tautomeric forms made it possible to exclude OH-tautomers of molecules E and F from further consideration.

Possible tautomers of molecule D were studied in detail by various quantum-chemical methods in the gas phase approximation [35]. The most energetically favorable neutral tautomer of D was shown to have the structure presented in Fig. **(2.8.d)**, while the overall contribution of other structures to tautomeric equilibrium did not exceed 0.02%. Similar results were obtained for other hydroxy-gamma-pyrones [35, 36], thus only the analogous structure of neutral acid A (Fig. **2.8.a**) will be considered herein.

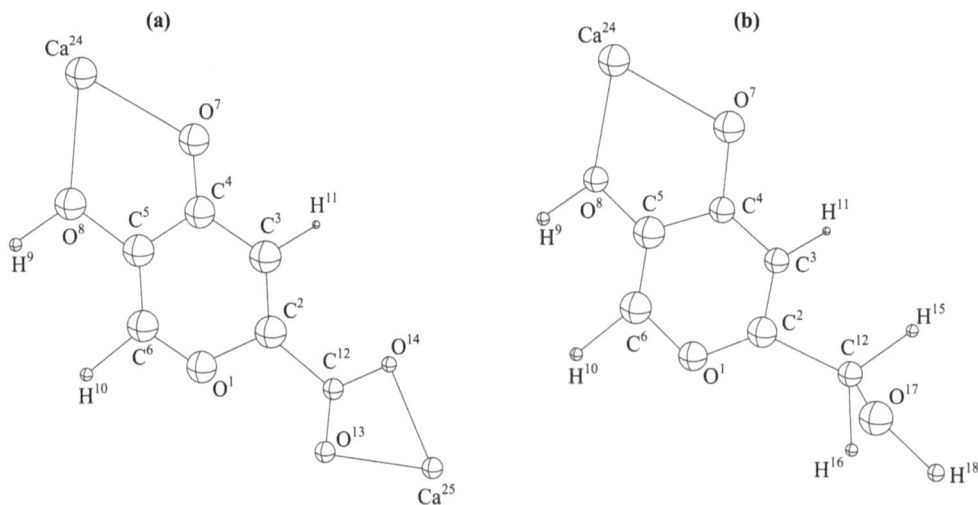

Fig. 3.8 contd.....

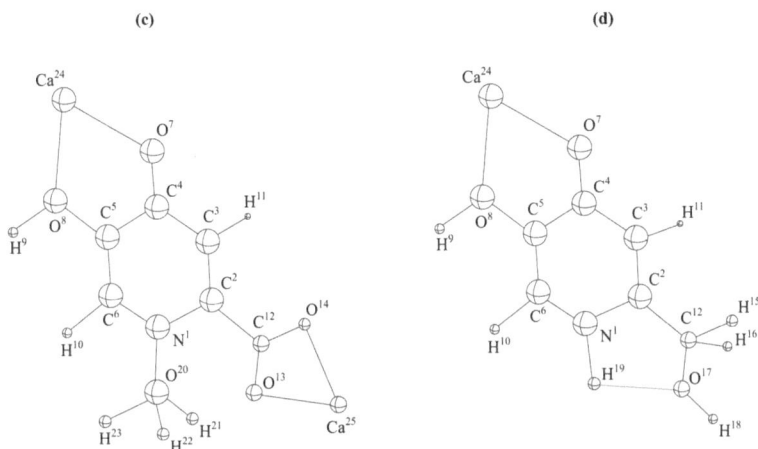

Fig. (3.8). Steric structures of gamma-pyrones and gamma-pyridones in the forms including the maximal amount of bound cations and accepted atom numbering.
a – calcium salt of calcium chelate complex of comenic acid (substance A);
b – calcium chelate complex of kojic acid (substance D);
c – calcium salt of calcium chelate complex of 5-hydroxy-1-methyl-gamma-pyridone-2-carboxylic acid (substance E);
d – calcium chelate complex of 5-hydroxy-2-hydroxymethyl-gamma-pyridone (substance F).
The dashed line represents the hydrogen bond.

Analysis of tautomeric equilibria in molecules A, D, E, and F indicates that only one among possible molecular forms of these compounds should be taken into account while performing quantum-chemical calculations to simulate the conditions of our patch-clamp experiments and interpreting their results. Fig. (**3.8**) displays steric structures of molecules A, D, E, and F with the maximal amount of bound cations and the atom numbering.

Free Acids

Selected bond lengths, bond and torsion angles, and Mulliken atomic charges in the molecules of free acids A, D, E, and F are collected in Table **3.1**. The heterocyclic fragments are almost planar due to the presence of conjugated double bonds (C^2—C^3, C^4—O^7, C^5—C^6). Deviation of the oxygen atoms in the carboxyl group of molecule E from the heteroring plane is caused by steric repulsion from the N-methyl substituent. The presence of this substituent is also responsible for elongation of N^1—C^2 bond, shortening of C^4—C^5 bond, increase of $C^2C^3C^4$ and $C^5C^6N^1$ bond angles, and decrease of $C^3C^4C^5$ and $C^6N^1C^2$ bond angles, as compared with the corresponding parameters of molecule F. No such differences were observed between the geometric parameters of molecules A and D. The hydroxymethyl groups in molecules D and F also deviate from the heteroring plane. Molecule F involves a weak intramolecular hydrogen bond N^1—$H^{19}\cdots O^{17}$, which affects $C^3C^2C^{12}O^{17}$ torsion angle and $C^2C^{12}O^{17}$ bond angle. All free acids A, D, E, and F are also characterized by a stronger intramolecular hydrogen bond O^8—$H^9\cdots O^7$.

Table 3.1. Selected calculated (RHF/6-31G* + PCM) and experimental bond lengths, bond and torsion angles, and atomic charges in the free acids A, D, E, and F.

Parameter[a]	A	D			E		F	
	$\varepsilon = 78.3$	$\varepsilon = 1$	$\varepsilon = 78.3$	Experiment [36]	$\varepsilon = 1$	$\varepsilon = 78.3$	$\varepsilon = 1$	$\varepsilon = 78.3$
Bond Length, *r*, Å								
X^1—C^2	1.327	1.329	1.331	1.352	1.361	1.352	1.347	1.339
C^2—C^3	1.333	1.336	1.339	1.343	1.352	1.357	1.352	1.360
C^3—C^4	1.454	1.452	1.446	1.431	1.445	1.435	1.443	1.429
C^4—C^5	1.462	1.469	1.464	1.445	1.453	1.443	1.467	1.456
C^5—C^6	1.330	1.326	1.328	1.336	1.338	1.343	1.335	1.339
X^1—C^6	1.346	1.352	1.348	1.360	1.375	1.366	1.377	1.367
C^4—O^7	1.211	1.206	1.216	1.243	1.214	1.229	1.215	1.233
C^5—O^8	1.345	1.341	1.347	1.347	1.342	1.350	1.342	1.352
C^2—C^{12}	1.501	1.501	1.501	1.502	1.502	1.505	1.512	1.514
C^{12}—O^{13}	1.189				1.188	1.190		
C^{12}—O^{14}	1.310				1.322	1.314		
C^{12}—O^{17}		1.397	1.402	1.404			1.401	1.400
N^1—C^{20}					1.467	1.474		
O^7—H^9	2.205	2.152	2.174		2.146	2.169	2.103	2.113
O^{17}—H^{19}							2.309	2.261
Bond Angle, ω, deg								
$X^1C^2C^3$	123.8	122.9	122.5	122.0	122.3	122.3	121.1	120.7
$C^2C^3C^4$	119.6	120.3	120.3	121.0	121.7	121.5	120.9	120.9
$C^3C^4C^5$	114.2	114.0	114.3	114.7	113.7	113.9	114.9	115.1
$C^4C^5C^6$	120.4	120.2	120.2	120.9	121.1	121.2	121.1	121.1
$C^5C^6X^1$	122.4	122.4	122.2	121.9	123.2	122.7	120.4	120.2
$C^6X^1C^2$	119.7	120.2	120.5	119.5	118.0	118.3	121.7	122.0
$C^4C^5O^8$	117.7	117.0	117.5		117.1	117.6	116.3	116.7
$C^5C^4O^7$	120.8	120.0	120.0		120.5	120.6	119.1	119.1

(Table 3.1) contd.....

Parameter[a]	A	D			E		F	
	$\varepsilon = 78.3$	$\varepsilon = 1$	$\varepsilon = 78.3$	Experiment [36]	$\varepsilon = 1$	$\varepsilon = 78.3$	$\varepsilon = 1$	$\varepsilon = 78.3$
$C^3C^2C^{12}$	121.7	125.0	125.2		118.9	118.6	123.4	123.3
$C^2C^{12}O^{13}$	122.1				124.8	124.0		
$O^{13}C^{12}O^{14}$	124.8				122.6	123.9		
$C^2C^{12}O^{17}$		108.2	107.9				112.2	111.9
$C^2N^1C^{20}$					124.5	124.0		
Torsion Angle, τ, deg								
$N^1C^2C^{12}O^{13}$					22.2	34.5		
$C^3C^2C^{12}O^{17}$		63.4	59.6				29.4	21.2
Atomic Charge, q, a.u.								
X^1	-0.61	-0.61	-0.61		-0.74	-0.73	-0.83	-0.82
C^2	0.34	0.45	0.44		0.32	0.36	0.36	0.38
C^3	-0.32	-0.39	-0.40		-0.36	-0.38	-0.41	-0.43
C^4	0.56	0.55	0.56		0.55	0.56	0.55	0.55
C^5	0.25	0.25	0.24		0.26	0.23	0.26	0.23
C^6	0.13	0.11	0.12		0.06	0.08	0.05	0.07
O^7	-0.67	-0.63	-0.70		-0.66	-0.75	-0.67	-0.77
O^8	-0.78	-0.76	-0.77		-0.77	-0.79	-0.77	-0.80
C^{12}	0.81	-0.03	-0.03		0.77	0.77	-0.02	-0.01
O^{13}	-0.58				-0.55	-0.58		
O^{14}	-0.69				-0.69	-0.69		
O^{17}		-0.73	-0.77				-0.74	-0.78
C^{20}					-0.27	-0.29		

[a] Molecules A and D, X = O; molecules E and F, X = N.

Comparison of the structural and electronic parameters of molecules E and F with those of A and D makes it possible to investigate the effect of the heteroatom on geometry of the molecules in study. Insofar as N—C bonds are usually longer than O—C bonds of the same multiplicity, X^1—C^2, C^2—C^3, C^5—C^6, and X^1—C^6 bonds in F and G are longer, while C^3—C^4 and C^4—C^5 bonds are shorter than the corresponding bonds in molecules A and D. Increase in electronegativity of the heteroatom in going from gamma-pyrones to gamma-pyridones is accompanied by redistribution of electron density on the neighboring carbon atoms, thus the negative charge on C^2, C^3, and C^6 atoms increases. The negative charge on O^7 atom also increases appreciably, and C^4—O^7 bond becomes longer; in addition, strengthening of O^8—$H^9 \cdots O^7$ intramolecular hydrogen bond is observed due to contribution of OH-forms to the resulting conformation of the free acids E and F.

Variation of dielectric constant ε (in other words, solvation of the molecules) affects the order of endocyclic bonds: N^1—C^2, C^3—C^4, C^4—C^5, and N^1—C^6 bonds in molecules E and F become stronger, while C^2—C^3 and C^5—C^6 bonds weaken; analogous effect on the geometric parameters of A and D is insignificant. The negative charge on O^7 and O^8 atoms increases under the effect of the outer field, which results in weakening of O^8—$H^9 \cdots O^7$ hydrogen bonds and C^4—O^7, C^5—O^8 bonds due to enhanced electrostatic repulsion between O^7 and O^8. The gain in the negative charge on O^{17} atom in molecule F is responsible for strengthening of N^1—$H^{19} \cdots O^{17}$ hydrogen bond. Almost no differences in the geometry and electronic parameters of the free acids A, D, E, and F at $\varepsilon = 10$ and $\varepsilon = 78.3$ were observed.

The optimized geometry of molecule D is very consistent with the experimental data [37] (Table **3.1**). The difference in the bond lengths does not exceed 0.02 Å, and the bond angles differ by no more than 1°. Although the Hartree-Fock approximation does not take into account electron correlation, consideration of solvation effects in terms of the PCM model ensures better accordance with the experimental data for molecule D, as compared to B3LYP/6-311++G(d,p) calculations in the gas phase [35]. The length of C^4—O^7 bond turned out to be somewhat underestimated in the framework of RHF/6-31G* + PCM approach; on the other hand, the applied procedure provides a fairly reliable data on the geometry and electronic structure of the free acids A, D, E, and F.

Complexes with Ca²⁺

As noted above, presence of the carbonyl and hydroxyl groups in the neighboring positions of the heteroring in molecules A, D, E, and F is favorable for chelation of divalent cations. Only a few studies were dedicated to investigation of the complex formation of hydroxypyridones and hydroxypyrones with Ca^{2+} [20, 38]; however, such complexes were shown to be stable in aqueous solution at physiological pH. In our patch-clamp experiments, the concentration of Ca^{2+} ions exceeded those of the agents applied by several orders of magnitude. Therefore, we can assert that the examined molecules form complexes with calcium in 1:1 stoichiometry. Selected bond lengths, bond and torsion angles, and atomic charges in calcium chelates of A, D, E, and F are given in Table **3.2**.

The effects of the heteroatom nature and substituent on the nitrogen atom on the geometry and electronic parameters of calcium chelates were essentially the same as in the series of free acids. Solvation of the complexes almost did not affect the structure of the heteroring and charges on the ring atoms; only the charge on C^5

increased slightly. Increase in the negative charges on O^{13}, O^{14}, and O^{17} atoms induced shortening of C^2—C^{12} bond and extension of C^{12}—O^{13} and C^{12}—O^{14} bonds. Correspondingly, changes in $C^3C^2C^{12}$, $C^2C^{12}O^{13}$, $C^2C^{12}O^{17}$, and $C^{13}C^{12}O^{14}$ bond angle values were observed. Shortening of C^5—O^8 and N^1—C^{20} bonds was appreciable. The positive charge on Ca^{24} atom increased by 0.02–0.03 a.u., and this cation shifted toward O^8, thus affecting $C^5C^4O^7$, $C^4O^7Ca^{24}$, $C^5O^8Ca^{24}$, and $O^7Ca^{24}O^8$ bond angle values.

Table 3.2. Selected bond lengths, bond angles, torsion angles, and atomic charges in calcium chelate complexes of A, D, E, and F obtained by full 6-31G*/RHF geometry optimization with varied dielectric constant ε.

Parameter[a]	A		D		E		F	
	$\varepsilon = 1$	$\varepsilon = 78.3$	$\varepsilon = 1$	$\varepsilon = 78.3$	$\varepsilon = 1$	$\varepsilon = 78.3$	$\varepsilon = 1$	$\varepsilon = 78.3$
Bond Length, r, Å								
X^1—C^2	1.324	1.324	1.326	1.326	1.356	1.355	1.338	1.334
C^2—C^3	1.346	1.346	1.354	1.346	1.356	1.360	1.372	1.367
C^3—C^4	1.422	1.434	1.415	1.423	1.408	1.414	1.400	1.404
C^4—C^5	1.436	1.435	1.438	1.443	1.412	1.414	1.430	1.435
C^5—C^6	1.339	1.317	1.335	1.331	1.354	1.352	1.346	1.344
X^1—C^6	1.317	1.337	1.318	1.327	1.344	1.347	1.349	1.353
C^4—O^7	1.249	1.253	1.254	1.243	1.268	1.263	1.269	1.266
C^5—O^8	1.388	1.388	1.390	1.370	1.394	1.372	1.395	1.378
C^2—C^{12}	1.516	1.496	1.506	1.500	1.519	1.511	1.513	1.510
C^{12}—O^{13}	1.177	1.186			1.179	1.187		
C^{12}—O^{14}	1.303	1.319			1.309	1.311		
C^{12}—O^{17}			1.386	1.398			1.390	1.398
N^1—C^{20}					1.497	1.485		
O^7—Ca^{24}	2.218	2.336	2.211	2.271	2.190	2.250	2.185	2.238
O^8—Ca^{24}	2.484	2.342	2.466	2.431	2.467	2.399	2.454	2.405
O^{17}—H^{19}							2.020	2.088
Bond Angle, ω, deg								
$X^1C^2C^3$	122.6	122.9	121.1	121.8	121.7	121.7	120.0	120.1
$C^2C^3C^4$	118.6	118.9	119.4	119.3	120.8	121.0	120.0	120.1
$C^3C^4C^5$	115.9	114.9	116.0	115.8	115.3	115.1	116.6	116.5
$C^4C^5C^6$	120.6	121.4	120.4	120.4	121.5	121.4	121.4	121.2
$C^5C^6X^1$	120.6	121.4	120.6	120.8	121.7	121.9	118.9	118.8
$C^6X^1C^2$	121.7	120.4	122.5	122.0	119.0	118.9	123.1	123.2

(Table 3.2) contd.....

Parameter[a]	A		D		E		F	
	$\varepsilon = 1$	$\varepsilon = 78.3$	$\varepsilon = 1$	$\varepsilon = 78.3$	$\varepsilon = 1$	$\varepsilon = 78.3$	$\varepsilon = 1$	$\varepsilon = 78.3$
$C^4C^5O^8$	114.0	113.0	114.3	113.9	114.5	114.4	114.3	114.3
$C^5C^4O^7$	119.0	121.0	118.7	119.9	119.8	120.7	118.6	119.4
$C^3C^2C^{12}$	121.7	122.4	126.3	126.0	119.2	118.2	123.6	122.8
$C^2C^{12}O^{13}$	119.5	122.1			122.9	124.0		
$O^{13}C^{12}O^{14}$	128.4	124.8			126.1	124.3		
$C^2C^{12}O^{17}$			105.7	106.5			106.8	108.3
$C^2N^1C^{20}$					123.7	123.8		
$C^4O^7Ca^{24}$	126.1	119.1	125.6	123.0	124.7	121.0	125.6	122.5
$C^5O^8Ca^{24}$	113.6	117.9	113.5	115.5	112.6	114.9	113.2	115.0
$O^7Ca^{24}O^8$	67.2	68.4	67.8	67.7	68.3	68.9	68.3	68.8
Torsion Angle, τ, deg								
$N^1C^2C^{12}O^{13}$					22.9	24.3		
$C^3C^2C^{12}O^{17}$			56.6	56.9			-0.2	-1.9
Atomic Charge, q, a.u.								
X^1	-0.54	-0.58	-0.53	-0.57	-0.72	-0.72	-0.81	-0.81
C^2	0.34	0.36	0.46	0.46	0.33	0.34	0.39	0.40
C^3	-0.30	-0.29	-0.36	-0.37	-0.31	-0.33	-0.37	-0.39
C^4	0.65	0.66	0.65	0.66	0.64	0.64	0.64	0.63
C^5	0.22	0.26	0.21	0.25	0.23	0.27	0.22	0.25
C^6	0.22	0.22	0.22	0.21	0.14	0.15	0.14	0.14
O^7	-0.84	-0.80	-0.85	-0.83	-0.88	-0.88	-0.89	-0.89
O^8	-0.89	-0.88	-0.90	-0.89	-0.90	-0.90	-0.90	-0.89
C^{12}	0.81	0.82	-0.02	-0.03	0.80	0.81	0.01	0.01
O^{13}	-0.48	-0.56			-0.49	-0.57		
O^{14}	-0.67	-0.70			-0.68	-0.69		
O^{17}			-0.72	-0.76			-0.77	-0.79
C^{20}					-0.30	-0.29		
Ca^{24}	1.77	1.80	1.76	1.79	1.75	1.77	1.74	1.77

[a] Molecules A and D, X = O; molecules E and F, X = N.

Due to the presence of electron-accepting calcium ion, the negative charges on O^7 and O^8 atoms increased by 0.12–0.22 and 0.09–0.13 a.u., respectively. Chelation-induced electron density redistribution resulted in an increase of charges on the ring carbon atoms, primarily, on C^4 and C^6. Complex formation enhances π-electron delocalization which leads to leveling of the endocyclic bond lengths. This effect is especially clearly observed in the calcium chelate of E: N^1—C^2, C^2—C^3, C^5—C^6, and N^1—C^6 bonds have almost similar lengths, *i.e.*, the ligand has nearly aromatic structure in the complex. Endocyclic bond leveling is more pronounced in calcium chelates of examined gamma-pyridones, which is consistent with a higher degree of π-electron density delocalization in these molecules, as compared to gamma-pyrone derivatives [39]. Analogous pattern was detected while studying hydroxypyridone chelates with several trivalent metal cations by X-ray diffraction [18, 19].

Obviously, chelation of calcium makes O^8—H^9···O^7 hydrogen bonding impossible; on the other hand, N^1—H^{19}···O^{17} hydrogen bond simultaneously becomes stronger. The distance $r(O^{17}$···$H^{19})$ decreases by ~0.2–0.3 Å, and $C^3C^2C^{12}O^{17}$ torsion angle in the calcium chelate of E approaches 0°, which is crucial for interpreting a probable mechanism of ligand-receptor binding of this gamma-pyridone derivative (see below).

Acid Anions

As it was already noted, deprotonation of the hydroxyl groups is not considered herein. The acidity constant of the carboxyl group of molecule A pK_a(COOH) equals to 1.4 [40]. No published data is available on the acidity of the carboxyl group in molecule E, but we presume that this acid is comparable in strength to comenic acid. Thus, only the carboxyl groups in A and E are ionized at physiological pH.

Differences in the geometry and electronic structure between the anions of A and E are essentially the same as those found between the free acids A and E (Table **3.3**). Structural changes induced by ionization of the carboxyl group were examined in detail in Chapter **2**. Upon ionization of isolated molecules, intramolecular hydrogen bonds O^8—H^9···O^7 become stronger, C^5—O^{12} bonds weaken, the negative charges on O^{12} increase in the absolute value, while $C^4C^5O^8$ and $C^5C^4O^7$ bond angle values decrease.

Table 3.3. Selected bond lengths, bond angles, torsion angles, and atomic charges in the anions of A and E obtained by full 6-31G*/RHF geometry optimization with varied dielectric constant ε.

Parameter[a]	A		E	
	$\varepsilon = 1$	$\varepsilon = 78.3$	$\varepsilon = 1$	$\varepsilon = 78.3$
Bond Length, r, Å				
X^1—C^2	1.335	1.332	1.359	1.349
C^2—C^3	1.347	1.340	1.365	1.364

(Table 3.3) contd.....

Parameter[a]	A		E	
	$\varepsilon = 1$	$\varepsilon = 78.3$	$\varepsilon = 1$	$\varepsilon = 78.3$
C^3—C^4	1.437	1.444	1.425	1.422
C^4—C^5	1.466	1.461	1.456	1.448
C^5—C^6	1.327	1.329	1.335	1.340
X^1—C^6	1.339	1.343	1.375	1.371
C^4—O^7	1.220	1.218	1.230	1.236
C^5—O^8	1.350	1.348	1.353	1.353
C^2—C^{12}	1.563	1.539	1.563	1.537
C^{12}—O^{13}	1.225	1.230	1.230	1.231
C^{12}—O^{14}	1.220	1.228	1.221	1.230
N^1—C^{20}			1.462	1.469
O^7—H^9	2.075	2.192	2.040	2.120
Bond Angle, ω, deg				
$X^1C^2C^3$	121.4	122.0	120.5	121.3
$C^2C^3C^4$	121.1	120.7	122.6	122.0
$C^3C^4C^5$	114.0	114.2	114.0	114.2
$C^4C^5C^6$	120.1	120.1	121.1	121.3
$C^5C^6X^1$	122.6	122.4	122.3	121.9
$C^6X^1C^2$	120.7	120.6	119.6	119.4
$C^4C^5O^8$	116.4	117.9	116.0	116.9
$C^5C^4O^7$	118.5	120.2	118.3	119.6
$C^3C^2C^{12}$	124.0	124.1	119.1	119.0
$C^2C^{12}O^{13}$	114.5	115.8	114.4	115.5
$O^{13}C^{12}O^{14}$	132.7	129.4	131.3	129.1
$C^2N^1C^{20}$			123.4	122.7
Torsion Angle, τ, deg				
$N^1C^2C^{12}O^{13}$			33.6	52.7
Atomic Charge, q, a.u.				
X^1	-0.58	-0.61	-0.72	-0.72
C^2	0.29	0.31	0.32	0.37
C^3	-0.39	-0.38	-0.42	-0.44
C^4	0.56	0.56	0.56	0.55
C^5	0.23	0.24	0.23	0.22
C^6	0.11	0.12	0.05	0.07

(*Table 3.3*) *contd.....*

Parameter[a]	A		E	
	$\varepsilon = 1$	$\varepsilon = 78.3$	$\varepsilon = 1$	$\varepsilon = 78.3$
O^7	-0.70	-0.71	-0.74	-0.79
O^8	-0.72	-0.79	-0.80	-0.80
C^{12}	0.76	0.75	0.75	0.71
O^{13}	-0.72	-0.76	-0.73	-0.75
O^{14}	-0.71	-0.76	-0.70	-0.76
C^{20}			-0.25	-0.28

[a] Molecule A, X = O; molecule E, X = N.

Calcium Salts

In accordance with ligand-receptor binding mechanism proposed by us for gamma-pyrones and gamma-pyridones, the counterion for the carboxyl group should be capable of forming a salt bond with a negatively charged functional group of the opioid-like receptor. Therefore, it was inappropriate to consider electroneutral salt molecules. Calcium salts of molecules A and E with 1:1 stoichiometry carry a single positive charge (+1) and differ from the corresponding neutral free acids only in the structure of the carboxyl groups, whereas no appreciable differences were observed in the bond lengths and bond angles within the heterocyclic fragment (Table **3.4**). The structures of the carboxyl groups in the calcium salts somewhat differ from the structures of those in the anions: C^2—C^{12} bonds are noticeably shorter, C^{12}—O^{13} and C^{12}—O^{14} bonds are longer, $C^3C^2C^{12}$ and $O^{13}C^{12}O^{14}$ bond angles are smaller, and $C^2C^{12}O^{13}$ bond angles are larger. This is the result of a growth of the negative charges on C^{12}, O^{13}, and O^{14} atoms due to salt formation. Solvation of the salt molecules makes C^{12}—O^{13} and C^{12}—O^{14} bonds stronger, while C^2—C^{12}, O^{13}—Ca^{25}, and O^{14}—Ca^{25} bonds weaken, $C^3C^2C^{12}$ and $O^{13}C^{12}O^{14}$ bond angles increase, $C^2C^{12}O^{13}$ angles become smaller, and the charges on O^7 and Ca^{25} grow in their absolute values.

Table 3.4. Selected bond lengths, bond angles, torsion angles, and atomic charges in calcium salts of A and E obtained by full 6-31G*/RHF geometry optimization with varied dielectric constant ε.

Parameter[a]	A		E	
	$\varepsilon = 1$	$\varepsilon = 78.3$	$\varepsilon = 1$	$\varepsilon = 78.3$
Bond Length, *r*, Å				
X^1—C^2	1.327	1.328	1.364	1.354
C^2—C^3	1.332	1.333	1.351	1.357
C^3—C^4	1.465	1.455	1.450	1.433
C^4—C^5	1.468	1.461	1.453	1.441

(Table 3.4) contd.....

Parameter[a]	A		E	
	$\varepsilon = 1$	$\varepsilon = 78.3$	$\varepsilon = 1$	$\varepsilon = 78.3$
C^5—C^6	1.330	1.330	1.340	1.343
X^1—C^6	1.353	1.345	1.377	1.366
C^4—O^7	1.200	1.211	1.209	1.230
C^5—O^8	1.335	1.345	1.338	1.351
C^2—C^{12}	1.490	1.501	1.494	1.509
C^{12}—O^{13}	1.257	1.247	1.258	1.246
C^{12}—O^{14}	1.251	1.245	1.258	1.248
N^1—C^{20}			1.465	1.475
O^{13}—Ca^{25}	2.264	2.334	2.269	2.332
O^{14}—Ca^{25}	2.264	2.334	2.248	2.323
O^7—H^9	2.210	2.198	2.187	2.166
Bond Angle, ω, deg				
$X^1C^2C^3$	124.4	123.7	122.4	122.0
$C^2C^3C^4$	119.4	119.6	121.9	121.9
$C^3C^4C^5$	113.9	114.2	113.5	113.8
$C^4C^5C^6$	120.4	120.4	121.0	121.2
$C^5C^6X^1$	122.7	122.4	123.5	122.8
$C^6X^1C^2$	119.2	119.7	117.5	118.2
$C^4C^5O^8$	117.5	117.6	117.5	117.6
$C^5C^4O^7$	121.3	120.7	121.3	120.7
$C^3C^2C^{12}$	122.4	123.1	116.4	117.4
$C^2C^{12}O^{13}$	120.3	119.4	121.8	120.5
$O^{13}C^{12}O^{14}$	120.4	122.1	119.1	121.2
$C^2N^1C^{20}$			124.8	124.5
Torsion Angle, τ, deg				
$N^1C^2C^{12}O^{13}$			17.1	25.0
Atomic Charge, q, a.u.				
X^1	-0.61	-0.61	-0.75	-0.73
C^2	0.31	0.33	0.32	0.35
C^3	-0.31	-0.32	-0.35	-0.38
C^4	0.53	0.56	0.55	0.56
C^5	0.27	0.25	0.27	0.23
C^6	0.10	0.13	0.06	0.08

(Table 3.4) contd.....

Parameter[a]	A		E	
	$\varepsilon = 1$	$\varepsilon = 78.3$	$\varepsilon = 1$	$\varepsilon = 78.3$
O^7	-0.58	-0.68	-0.63	-0.76
O^8	-0.75	-0.78	-0.75	-0.79
C^{12}	0.88	0.89	0.88	0.87
O^{13}	-0.79	-0.79	-0.80	-0.78
O^{14}	-0.77	-0.78	-0.79	-0.79
C^{20}			-0.27	-0.28
Ca^{25}	1.65	1.73	1.65	1.73

[a] Molecule A, X = O; molecule E, X = N.

Calcium Salts of Ca^{2+} Complexes

At physiological pH only calcium complexes of molecules A and E are obviously capable of forming salts with Ca^{2+} ions, the charge of these salts being equal to +3. Chelation of Ca^{2+} makes the heteroring considerably more rigid; therefore, almost no differences in the geometry and electronic structure of heterocycles in the molecules of calcium chelates of A and E and their calcium salts were detected. Selected bond lengths, bond angles, torsion angles, and atomic charges in calcium salts of calcium chelates of A and E are given in Table **3.5**. Comparison of the data presented in Tables **3.4** and **3.5** demonstrates that chelation of Ca^{2+} somewhat changes the structure of the carboxyl group: elongation of C^2—C^{12}, O^{13}—Ca^{25}, O^{14}—Ca^{25} bonds and shortening of C^{12}—O^{13}, C^{12}—O^{14} bonds is observed, and the charges on C^{12}, O^{13}, O^{14}, and Ca^{25} atoms increase relative to those in the corresponding calcium salts. Solvation of the chelate salts almost does not affect the structure of the heterocyclic fragment, and variations are detected only in the geometric parameters of the carboxyl groups and Ca^{24} atom position.

Table 3.5. Selected bond lengths, bond angles, torsion angles, and atomic charges in calcium salts of calcium chelate complexes of A and E obtained by full 6-31G*/RHF geometry optimization with varied dielectric constant ε.

Parameter[a]	A		E	
	$\varepsilon = 1$	$\varepsilon = 78.3$	$\varepsilon = 1$	$\varepsilon = 78.3$
Bond Length, r, Å				
X^1—C^2	1.326	1.326	1.362	1.356
C^2—C^3	1.343	1.338	1.363	1.357
C^3—C^4	1.433	1.436	1.416	1.419
C^4—C^5	1.435	1.434	1.411	1.416

(Table 3.5) contd.....

Parameter[a]	A		E	
	$\varepsilon = 1$	$\varepsilon = 78.3$	$\varepsilon = 1$	$\varepsilon = 78.3$
$C^5—C^6$	1.340	1.334	1.356	1.353
$X^1—C^6$	1.319	1.324	1.346	1.347
$C^4—O^7$	1.244	1.240	1.263	1.263
$C^5—O^8$	1.383	1.368	1.388	1.372
$C^2—C^{12}$	1.516	1.505	1.523	1.512
$C^{12}—O^{13}$	1.242	1.242	1.244	1.243
$C^{12}—O^{14}$	1.236	1.242	1.244	1.247
$N^1—C^{20}$			1.495	1.484
$O^7—Ca^{24}$	2.246	2.309	2.211	2.248
$O^8—Ca^{24}$	2.509	2.382	2.498	2.381
$O^{13}—Ca^{25}$	2.354	2.346	2.346	2.347
$O^{14}—Ca^{25}$	2.341	2.343	2.329	2.335
Bond Angle, ω, deg				
$X^1C^2C^3$	122.7	122.9	121.4	122.1
$C^2C^3C^4$	118.6	119.1	121.2	120.8
$C^3C^4C^5$	115.7	115.2	115.1	114.8
$C^4C^5C^6$	120.6	120.7	121.4	121.6
$C^5C^6X^1$	120.8	121.4	122.1	121.9
$C^6X^1C^2$	121.6	120.8	118.8	118.6
$C^4C^5O^8$	113.8	114.1	114.3	114.7
$C^5C^4O^7$	119.3	119.9	120.0	120.4
$C^3C^2C^{12}$	124.0	124.1	116.9	117.3
$C^2C^{12}O^{13}$	119.1	118.9	121.5	120.6
$O^{13}C^{12}O^{14}$	123.1	123.0	121.2	121.9
$C^2N^1C^{20}$			123.7	123.8
$C^4O^7Ca^{24}$	126.4	121.0	125.3	120.7
$C^5O^8Ca^{24}$	114.3	116.6	113.1	114.8
$O^7Ca^{24}O^8$	66.3	67.8	67.3	69.3
Torsion Angle, τ, deg				
$N^1C^2C^{12}O^{13}$			1.0	29.6
Atomic Charge, q, a.u.				
X^1	-0.54	-0.57	-0.73	-0.72
C^2	0.33	0.35	0.30	0.37

(Table 3.5) contd.....

Parameter[a]	A		E	
	$\varepsilon = 1$	$\varepsilon = 78.3$	$\varepsilon = 1$	$\varepsilon = 78.3$
C^3	-0.29	-0.29	-0.31	-0.33
C^4	0.65	0.65	0.64	0.64
C^5	0.23	0.26	0.24	0.27
C^6	0.22	0.21	0.14	0.14
O^7	-0.82	-0.81	-0.87	-0.88
O^8	-0.88	-0.88	-0.89	-0.90
C^{12}	0.88	0.90	0.91	0.87
O^{13}	-0.75	-0.77	-0.77	-0.77
O^{14}	-0.73	-0.77	-0.76	-0.77
C^{20}			-0.29	-0.29
Ca^{24}	1.78	1.80	1.76	1.77
Ca^{25}	1.73	1.74	1.73	1.74

[a] Molecule A, X = O; molecule E, X = N.

The distance between the calcium ions Ca^{24} and Ca^{25} ranges from 9.4 to 9.6 Å. We suppose that these cations form ion-ionic bonds with negatively charged functional groups in binding pocket of the opioid-like receptor. It is important that $r(Ca^{24}\cdots Ca^{25})$ depends only slightly on dielectric constant ε and this distance is practically the same in all examined molecules, which is required to provide the ligand complementarity to its receptor binding site.

Energy Effects of Salt Formation and Ca^{2+} Chelation

Table **3.6** contains the energies of formation of calcium chelates, calcium salts, and calcium chelates of calcium salts of molecules A, D, E, and F calculated according to the scheme described in Chapter **2** at different ε. The calculations were carried out both with and without account taken of the cavitation energy [41], as well as of the contributions of dispersion and exchange-repulsion forces [42] to the Gibbs energy of solvation. The energies of formation of calcium salts of A and E are almost similar. Chelation of Ca^{2+} and formation of calcium salts of the chelates is slightly more energetically favorable for gamma-pyridones E and F than for gamma-pyrones A and D, because O^7 and O^8 atoms in the former two molecules are charged more negatively than in the latter two (Table **3.1**).

Consideration of the cavitation energy and contributions of dispersion and exchange repulsion forces to the Gibbs energy of solvation decreases the energies of salt and complex formation by ~2.5–3.5 kcal·mol^{-1}, and the energies of chelate salt formation by ~6 kcal·mol^{-1} in the absolute value. As it might be expected, the energies of salt formation and Ca^{2+} chelation noticeably decrease upon solvation, but these processes remain nevertheless energetically favorable.

Table 3.6. Energies of formation of calcium salts (ΔE_{salt}(Ca)), calcium chelate complexes ($\Delta E_{complex}$(Ca)) and calcium salts of calcium chelate complexes ($\Delta E_{complex\text{-}salt}$ (Ca, Ca)) of A, D, E, and F with varied dielectric constant ε.

Parameter	ε	A	D	E	F
ΔE_{salt}(Ca), kcal·mol^{-1}	1	-283.6		-281.0	
	10	-89.0 (-92.0)[a]		-89.0 (-91.5)	
	78.3	-70.6 (-73.7)		-70.9 (-73.4)	
$\Delta E_{complex}$ (Ca), kcal·mol^{-1}	1	-105.5	-114.9	-125.8	-132.3
	10	-56.4 (-59.7)	-60.4 (-63.9)	-66.3 (-69.8)	-69.1 (-72.8)
	78.3	-50.4 (-53.0)	-55.7 (-59.2)	-61.3 (-64.8)	-62.9 (-66.6)
$\Delta E_{complex\text{-}salt}$ (Ca, Ca), kcal·mol^{-1}	1	-318.3		-335.5	
	10	-142.6 (-148.6)		-151.3 (-157.1)	
	78.3	-126.6 (-132.7)		-133.2 (-139.0)	

[a] Energies calculated without dispersion and exchange repulsion contributions are given in parentheses.

Quantum-chemical calculations demonstrate that substances A, D, E, and F in all examined molecular forms are characterized by essentially similar geometry and electronic structure. Some observed differences are related mainly to the nature of heteroatom in the ring: the endocyclic bond lengths in molecules E and F are leveled to a greater extent than those in molecules A and D, which indicates a higher degree of π-electron density delocalization in gamma-pyridone derivatives as compared to gamma-pyrones and hence a higher aromaticity of the former.

Structure of Molecule F (5-hydroxy-2-hydroxymethyl-gamma-pyridone)

In Chapter **2** we have put forward a suggestion that gamma-pyrones interact with the opioid-like receptor in the form of calcium salt of calcium chelate complex and that the ligand molecules should possess a number of pharmacophoric groups (the carboxyl group in position 2 and the hydroxyl or methoxy group in position 5 of the heterocyclic ring) in order to produce the physiological effect, *i.e.*, to modulate the functioning of $Na_V1.8$ channels. This assumption is consistent with the modulating effect of substance E which matches the above requirements. It is somewhat surprising that gamma-pyridone F was shown experimentally to exhibit channel-modulating activity despite the lack of the carboxyl group responsible for forming an ion-ionic bond with Ca^{2+} cation, while its gamma-pyrone analog D showed no such effect (Chapter **2**).

Molecules D and F are capable of binding only one Ca^{2+} ion in water solution by chelation involving the carbonyl and hydroxyl groups. Receptor binding pocket may be regarded as a heterogeneous anisotropic medium characterized by a relatively low dielectric constant ε, but appreciably higher polarizability as compared to water and hence a greater ability to stabilize structures unstable in aqueous environment. Probably, the ligand transfer from solution to receptor binding pocket upon formation of the ligand-receptor complex is accompanied by deprotonation of the hydroxymethyl group in position 2 of the heteroring in molecules D and F.

We examined model calcium salts of calcium chelates of D and F, which were constructed by replacement of H^{18} atom (see Fig. **3.8**) in the corresponding calcium chelates with a calcium atom. Various possible conformations of the hydroxymethyl groups were analyzed: $C^3C^2C^{12}O^{17}$ torsion angle in the model structures was varied with an increment of 30°, and full *ab initio* optimization of the geometric parameters was carried out at every step with $\varepsilon = 1$, 10 and 78.3 (RHF/6-31G* + PCM). The same conformation of each model compound was found to have the lowest energy at all ε values (Fig. **3.9**). The hydroxymethyl group in calcium salt of calcium chelate of D adopts the extended conformation in which $C^3C^2C^{12}O^{17}$ torsion angle is ~0° and $r(Ca^{24}\cdots Ca^{25})$ ranges from 8.6 to 9.2 Å. Analogous conformation of the hydroxymethyl group was detected while studying the structure of the anion of D by a number of calculational methods [29]. Intramolecular hydrogen bond N^1—$H^{19}\cdots O^{17}$ in calcium salt of calcium chelate of G fixates the hydroxymethyl group in the conformation with $C^3C^2C^{12}O^{17}$ torsion angle ~180° and $r(Ca^{24}\cdots Ca^{25})$ in the range of 9.9-10.0 Å.

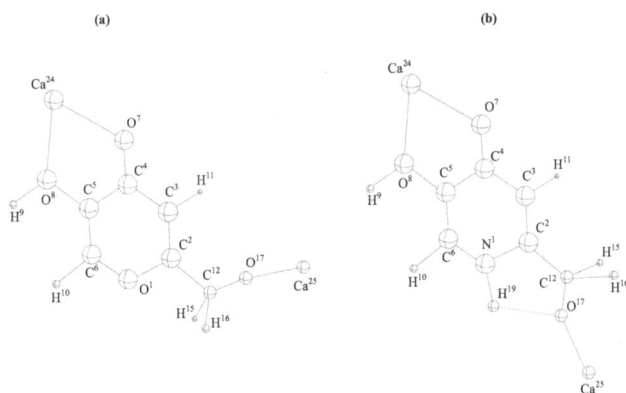

Fig. (3.9). Steric structures of model calcium salts of calcium complexes.
a – calcium salt of calcium chelate complex of kojic acid (substance D);
b – calcium salt of calcium chelate complex of 5-hydroxy-2-hydroxymethyl-gamma-pyridone (substance F).
The dashed line represents the hydrogen bond.

Comparison of steric structures of salts of A, E, and F calcium chelates shows insignificant difference in positions of calcium ions with respect to the heteroring, and intercationic distances depend on ε only slightly. Position of Ca^{25} atom in calcium salt of calcium chelate of D is appreciably different, and $r(Ca^{24}\cdots Ca^{25})$ distance varies over a considerably broad range depending on ε. Thus, the nature of the ring heteroatom is demonstrated to affect conformation of the hydroxymethyl group and determine the ability of calcium salts of D and F calcium chelates to bind to the opioid-like receptor. The lack of activity of substance D in our patch-clamp experiments is accounted for by the fact that the positions of Ca^{2+} ions in calcium salt of calcium chelate of this agent do not match the positions of nucleophilic sites in receptor binding pocket.

CONCLUSION

Summarizing the results of quantum-chemical calculations discussed in Chapters **2** and **3**, a set of structural criteria that determines the possibility for formation of ligand-receptor complexes between gamma-pyridones or gamma-pyrones and the opioid-like receptor is suggested: (1) in position 5 of the heterocycle should be present a hydroxyl or methoxy group which is capable, in combination with the carbonyl group in position 4, to chelate Ca^{2+} cation; (2) the second Ca^{2+} cation serves as the counterion at the deprotonated carboxyl or hydroxymethyl group in position 2 of the heterocycle; (3) intercationic distance $r(Ca^{2+}\cdots Ca^{2+})$ may range from 9.4 to 10.0 Å; and (4) Ca^{2+} cations should occupy specific positions with respect to the heterocycle. It is necessary to stress out that the major contribution to the energy of ligand-receptor binding of gamma-pyrones and gamma-pyridones is provided by strong ion-ionic interactions between bound calcium cations of the ligand and negatively charged aspartate residues of the opioid-like receptor. It is appropriate to use the structure of calcium salt of calcium chelate of comenic acid (substance A) as a reference, taking into account that comenic acid is capable of binding to the opioid-like receptor at the lowest concentration among examined substances.

While suggesting a possible mechanism of ligand-receptor binding for gamma-pyrones and gamma-pyridones, we should keep in mind that the physiologically active ligand form may be a structure unstable in bulk aqueous solution but capable of being stabilized *via* additional inter- and intramolecular interactions in heterogenous anisotropic conditions inherent to opioid-like receptor binding pocket. The presence or absence of receptor-coupled modulation of $Na_V 1.8$ sodium channels by examined agents cannot be predicted solely on the basis of their structural formulae, and detailed examination of their geometry and electronic parameters is absolutely required.

Channel-modulating activity of gamma-pyridones is expressed at higher concentrations than that of gamma-pyrones, which makes the former substances less effective. We found two feasible explanations for this fact: 1) both gamma-pyridones and gamma-pyrones studied are rather small molecules, therefore almost all their potentially functional atoms should be optimally involved in ligand-receptor complex formation, and even the slightest change in their structure should influence the energy of their binding to the opioid-like receptor; 2) existence of two tautomeric forms of gamma-pyridones results in an increase of the agent concentration required to obtain the effect comparable to that of gamma-pyrones, which adopt just one molecular form at all values of ε.

In the following chapter we report the results of combined calculational and patch-clamp study on several molecules that belong to a completely different structural class of naturally occurring steroids, and demonstrate that their physiological role in nociception can also be explained by taking into consideration possible involvement of Ca^{2+}.

CONFLICT OF INTEREST

The authors confirm that they have no conflict of interest to declare for this publication.

ACKNOWLEDGEMENTS

Declared none.

REFERENCES

[1]   Kendall DA, Browner M, Enna SJ. Comparison of the antinociceptive effect of gamma-aminobutyric acid (GABA) agonists: evidence for a cholinergic involvement. J Pharmacol Exp Ther 1982; 220(3): 482-7.
        [PMID: 7062260]

[2]   Aytemir MD, Uzbay T, Erol DD. New 4(1H)-pyridinone derivatives as analgesic agents. Arzneim-Forsch/Drug Res 1999; 49(3): 250-4.
        [PMID: 10219469]

[3]   Ozturk G, Erol DD, Uzbay T, Aytemir MD. Synthesis of 4(1H)-pyridinone derivatives and investigation of analgesic and antiinflammatory activities. Farmaco 2001; 56(4): 251-6.
        [http://dx.doi.org/10.1016/S0014-827X(01)01083-7] [PMID: 11421252]

[4]   Ozturk G, Erol DD, Aytemir MD, Uzbay T. New analgesic and antiinflammatory agents 4(1H)-pyridinone derivatives. Eur J Med Chem 2002; 37(10): 829-34.
        [http://dx.doi.org/10.1016/S0223-5234(02)01390-9] [PMID: 12446041]

[5]   Hajhashemi V, Saghaei L, Fassihi A, Mojiri-Froshani H. A study on the analgesic effects of four new derivatives of 3-hydroxy pyridine-4-one. Res Pharm Sci 2012; 7(1): 37-42.
        [PMID: 23181078]

[6]   Hajhashemi V, Mojiri-Froshani H, Saghaei L, Fassihi A. A study on the anti-inflammatory effects of new derivatives of 3-hydroxy pyridine-4-one. Adv Biomed Res 2014; 3: 134-9.

[http://dx.doi.org/10.4103/2277-9175.133276] [PMID: 24949305]

[7]   Timeus F, Valle P, Crescenzio N, *et al.* Effect of desferrioxamine and hydroxypyridones on hemopoietic progenitors and neuroectodermal tumor cells. Am J Hematol 1994; 47(3): 183-8.
[http://dx.doi.org/10.1002/ajh.2830470307] [PMID: 7524314]

[8]   Hwang DR, Proctor GR, Driscoll JS. Pyridones as potential antitumor agents II: 4-pyridones and bioisosteres of 3-acetoxy-2-pyridone. J Pharm Sci 1980; 69(9): 1074-6.
[http://dx.doi.org/10.1002/jps.2600690923] [PMID: 7411412]

[9]   Sakagami K, Iwamatsu K, Atsumi K, Hatanaka M. Synthetic cephalosporins. VII. Synthesis and antibacterial activity of 7-[(Z)-2-(2-aminothiazol-4-yl)-2-(3-(3-hydroxy-4-pyridon-1-yl)-3-carbo-xypropoxyimino)acetamido]-3-(1,2,3-thiadiazol-5-yl)-thiomethyl-3-cephem-4-carboxylic acid and its related compounds. Chem Pharm Bull (Tokyo) 1990; 38(12): 3476-9.
[http://dx.doi.org/10.1248/cpb.38.3476] [PMID: 2092946]

[10]   Feng MH, van der Does L, Bantjes A. Iron (III)-chelating resins. 3. Synthesis, iron (III)-chelating properties, and *in vitro* antibacterial activity of compounds containing 3-hydroxy-2-methyl-4 (1H)-pyridinone ligands. J Med Chem 1993; 36(19): 2822-7.
[http://dx.doi.org/10.1021/jm00071a013] [PMID: 8410996]

[11]   Erol DD, Yulug N. Synthesis and antimicrobial investigation of thiazolinoalkyl-4(1H)-pyridones. Eur J Med Chem 1994; 29(11): 893-7.
[http://dx.doi.org/10.1016/0223-5234(94)90113-9]

[12]   Hershko C, Theanacho EN, Spira DT, Peter HH, Dobbin P, Hider RC. The effect of N-alkyl modification on the antimalarial activity of 3-hydroxypyridin-4-one oral iron chelators. Blood 1991; 77(3): 637-43.
[PMID: 1991172]

[13]   Williams WR. Synthesis of some 4-pyranones and 4-pyridones structurally related to isoproterenol. Can J Chem 1976; 54(21): 3377-82.
[http://dx.doi.org/10.1139/v76-485]

[14]   Dexter DT, Carter CJ, Wells FR, *et al.* Basal lipid peroxidation in substantia nigra is increased in Parkinsons disease. J Neurochem 1989; 52(2): 381-9.
[http://dx.doi.org/10.1111/j.1471-4159.1989.tb09133.x] [PMID: 2911023]

[15]   Ma Y, Luo W, Quinn PJ, Liu Z, Hider RC. Design, synthesis, physicochemical properties, and evaluation of novel iron chelators with fluorescent sensors. J Med Chem 2004; 47(25): 6349-62.
[http://dx.doi.org/10.1021/jm049751s] [PMID: 15566304]

[16]   Kalinowski DS, Richardson DR. The evolution of iron chelators for the treatment of iron overload disease and cancer. Pharmacol Rev 2005; 57(4): 547-83.
[http://dx.doi.org/10.1124/pr.57.4.2] [PMID: 16382108]

[17]   Thompson KH, Barta CA, Orvig C. Metal complexes of maltol and close analogues in medicinal inorganic chemistry. Chem Soc Rev 2006; 35(6): 545-56.
[http://dx.doi.org/10.1039/b416256k] [PMID: 16729148]

[18]   Scarrow RC, Riley PE, Abu-Dari K, White D, Raymond KN. Ferric ion sequestering agents. 13. Synthesis, structures, and thermodynamics of complexation of cobalt (III) and iron (III) tris complexes of several chelating hydroxypyridinones. Inorg Chem 1985; 24(6): 954-67.
[http://dx.doi.org/10.1021/ic00200a030]

[19]   Nelson WO, Karpishin TB, Rettig SJ, Orvig C. Aluminum and gallium compounds of 3-hydroxy-4-pyridinones: synthesis, characterization, and crystallography of biologically active complexes with unusual hydrogen bonding. Inorg Chem 1988; 27(6): 1045-51.
[http://dx.doi.org/10.1021/ic00279a022]

[20]   Stunzi H, Harris RL, Perrin DD, Teitei T. Stability constants for metal complexation by isomers of mimosine and related compounds. Aust J Chem 1980; 33(10): 2207-20.

[http://dx.doi.org/10.1071/CH9802207]

[21]   Hariharan PC, Pople JA. The influence of polarization functions on molecular orbital hydrogenation energies. Theor Chim Acta 1973; 28(3): 213-22.
[http://dx.doi.org/10.1007/BF00533485]

[22]   Schmidt MW, Baldridge KK, Boatz JA, *et al.* General atomic and molecular electronic structure system. J Comput Chem 1993; 14(11): 1347-63.
[http://dx.doi.org/10.1002/jcc.540141112]

[23]   Tomasi J, Persico M. Molecular interactions in solution: an overview of methods based on continuous distributions of the solvent. Chem Rev 1994; 94(7): 2027-94.
[http://dx.doi.org/10.1021/cr00031a013]

[24]   Petrola R. Stability of yttrium (iii) complexes of substituted 3-hydroxy-4H-pyran-4-ones in aqueous solution. Finn Chem Lett 1986; 13(5): 129-35.

[25]   Petrola R. Spectrophotometric study on the equilibrium of substituted 3-hydroxy-4H-pyran-4-ones with Zn(II) ions in aqueous solution. Finn Chem Lett 1985; 12(5): 219-24.

[26]   Imafuku K, Ishizaka M, Matsumura H. Substituent effects on 6-substituted 3-hydroxy-1-methyl-4-pyridones. Bull Chem Soc Jpn 1979; 52(1): 111-3.
[http://dx.doi.org/10.1246/bcsj.52.111]

[27]   Imafuku K, Ishizaka M, Matsumura H. Structure of 5-hydroxy-2-hydroxymethyl-4-pyridones. Bull Chem Soc Jpn 1979; 52(1): 107-10.
[http://dx.doi.org/10.1246/bcsj.52.107]

[28]   Luca C, Popescu DO, Constantinescu T. The amphionic structure of 3-hydroxy-2-methyl-4H-pyran-4-one and the properties of its complexes with iron ions. Rev Roum Chim 1993; 38(1): 123.

[29]   Albert A, Phillips JN. 264. Ionization constants of heterocyclic substances. Part II. Hydroxy-derivatives of nitrogenous six-membered ring-compounds. J Chem Soc 1956; 1294-304.
[http://dx.doi.org/10.1039/jr9560001294]

[30]   Gordon A, Katritzky AR, Roy SK. Tautomeric pyridines. Part X. Effects of substituents on pyridine-hydroxypyridine equilibria and pyridone basicity. J Chem Soc B 1968; 556-61.
[http://dx.doi.org/10.1039/J29680000556]

[31]   Besso H, Imafuku K, Matsumura H. Tautomerism of 4-pyridones. Bull Chem Soc Jpn 1977; 50(3): 710-2.
[http://dx.doi.org/10.1246/bcsj.50.710]

[32]   Schlegel HB, Gund P, Fluder EM. Tautomerization of formamide, 2-pyridone, and 4-pyridone: an ab initio study. J Am Chem Soc 1982; 104(20): 5347-51.
[http://dx.doi.org/10.1021/ja00384a017]

[33]   Molenda JJ, Jones MM, Basinger MA. Enhancement of iron excretion *via* monoanionic 3-hydroxypyrid-4-ones. J Med Chem 1994; 37(1): 93-8.
[http://dx.doi.org/10.1021/jm00027a011] [PMID: 8289206]

[34]   Zhang Z, Rettig SJ, Orvig C. Physical and structural studies of N-carboxymethyl- and N-(p-methoxyphenyl)-3-hydroxy-2-methyl-4-pyridinone. Can J Chem 1992; 70(3): 763-70.
[http://dx.doi.org/10.1139/v92-101]

[35]   Zborowski K, Grybos R, Proniewicz LM. Determination of the most stable structures of selected hydroxypyrones and their cations and anions. J Mol Struct THEOCHEM 2003; 639(1): 87-100.
[http://dx.doi.org/10.1016/S0166-1280(03)00586-4]

[36]   Zborowski K, Korenova A, Uher M, Proniewicz LM. Quantum chemical studies on tautomeric equilibria in chlorokojic and azidokojic acids. J Mol Struct THEOCHEM 2004; 683(1–3): 15-22.
[http://dx.doi.org/10.1016/j.theochem.2004.06.007]

[37]   Lokaj J, Kožíšek J, Koreň B, Uher M, Vrábel V. Structure of kojic acid. Acta Cryst 1991; C47(1): 193-4.

[38]   Okač A, Kolařik Z. Potentiometrische Untersuchung von Komplexsalzen der Kojisäure in wässrigen Lösungen. Collect Czech Chem Commun 1959; 24(1): 266-72.
        [http://dx.doi.org/10.1135/cccc19590266]

[39]   Nelson WO, Karpishin TB, Rettig SJ, Orvig C. Physical and structural studies of N-substituted-3-hydroxy-2-methyl-4(1H)-pyridinones. Can J Chem 1988; 66(1): 123-31.
        [http://dx.doi.org/10.1139/v88-019]

[40]   Petrola R. Spectrophotometric study on the equilibria of pyromeconic acid derivatives with proton in aqueous solution. Finn Chem Lett 1985; 12(5): 207-12.

[41]   Langlet J, Claverie P, Caillet J, Pullman A. Improvements of the continuum model. 1. Application to the calculation of the vaporization thermodynamic quantities of nonassociated liquids. J Phys Chem 1988; 92(6): 1617-31.
        [http://dx.doi.org/10.1021/j100317a048]

[42]   Amovilli C, Mennucci B. Self-consistent-field calculation of Pauli repulsion and dispersion contributions to the solvation free energy in the polarizable continuum model. J Phys Chem B 1997; 101(6): 1051-7.
        [http://dx.doi.org/10.1021/jp9621991]

CHAPTER 4

Possible Mechanisms of Ligand-Receptor Binding of Cardiotonic Steroids

Abstract: Cardiotonic steroids are a recently discovered class of hormones synthesized in the adrenal cortex and hypothalamus and circulating in the blood. It is well known that the target molecule for these agents is Na^+,K^+-ATPase. A direct consequence of the proposed mechanism of $Na_V1.8$ channels modulation in Chapter **1** is the prediction of a special signaling function of the sodium pump. In other words, Na^+,K^+-ATPase should be involved in the processing of nociceptive information. The data presented in the current chapter support this idea. According to our findings, ouabain as a newly recognized hormone plays the role of endogenous analgesic at subnanomolar concentrations. Its target site is located directly on Na^+,K^+-ATPase and it recognizes ouabain, only in the form of its calcium chelate complex. The most significant result discussed in this chapter is explanation of the dual effect of ouabain: two distinct attacking molecules (ouabain and its calcium chelate complex) bind to two distinct sites of Na^+,K^+-ATPase, thus modulating two distinct functions of the enzyme: pumping and non-pumping (signal-transducing). Another newly recognized hormone, marinobufagenin, also exhibits analgesic effect at low concentrations but it is of principally different nature. This molecule lacks the ability to form marinobufagenin–Ca^{2+} chelate complex in 1:1 stoichiometry which could activate the signal-transducing function of Na^+,K^+-ATPase upon binding to the enzyme. The decrease of Z_{eff} of $Na_V1.8$ channel activation gating device induced by application of marinobufagenin at nanomolar concentrations results from activation of the "modulated receptor" mechanism, *i.e.*, this molecule binds directly to the aminoacid sequence of the channel without involvement of Ca^{2+}.

Keywords: Ca^{2+} chelate complex, Limiting slope procedure, Marinobufagenin, $Na_V1.8$ channels, Na^+,K^+-ATPase, Nociception, Ouabagenin, Ouabain, Patch-clamp method, Quantum-chemical calculations.

Endogenous cardiotonic steroids have various physiological functions. In particular, it was demonstrated that abnormal concentrations of these agents [1] could evoke different pathological states: congestive heart failure, cardiac arrhythmias [2, 3], hypertension [4], cancer [5], and depressive disorders [6, 7]. An increase of concentration of cardiotonic steroids was detected in the blood and subcutaneous water upon stress, lassitude, inflammatory processes in the

Boris V. Krylov, Ilia V. Rogachevskii, Tatiana N. Shelykh, Vera B. Plakhova

organism [8], pregnancy [9], and as a result of nephrectomy [10]. Endogenous cardiotonic steroids also influence cell growth and proliferation [11 - 13]. Ouabain exhibits anti-apoptosis action on endothelial cells [14]. Bufadienolides may induce apoptosis in human leukemia cells [15] and they also display antiproliferative activity and immunosuppressive activity upon action on T cells [16].

PATCH-CLAMP INVESTIGATION OF CARDIOTONIC STEROIDS

A direct consequence of our working hypothesis (Fig. **1.17**) is the prediction of a new mechanism of action of cardiotonic steroids, which should play an important role in nociception. When these agents activate the transducing function of Na^+, K^+-ATPase, they might produce an analgesic effect. In our opinion, the most interesting objects to study are endogenous substances, such as ouabain and marinobufagenin. Obviously, these agents can exert their analgesic properties only in vanishingly small "endogenous" concentrations, since it is known that high concentrations of cardiotonic steroids are extremely toxic. Endogenous ouabain was found in blood plasma in subnanomolar concentrations [17 - 19]. The designation "ouabain–Ca^{2+}" is used further to distinguish low (endogenous) concentrations of ouabain from its high concentrations, as in physiologically adequate conditions endogenous ouabain should exist in the form of calcium chelate complex (see below). Fig. (**4.1**) illustrates $Na_V1.8$ currents in the control experiment and after extracellular application of ouabain–Ca^{2+} at 1 nM. It is clearly seen that the amplitude values of the currents are decreased (Fig. **4.1.a**). The peak current-voltage curve shifts in the depolarizing direction (Fig. **4.1.b**) and the left branch of this function becomes steeper at negative E after ouabain–Ca^{2+} has been applied, which results in a very pronounced decrease in Z_{eff} (Fig. **4.2.a**) due to activation of the transducer-coupled membrane mechanism described in Chapter 1 (Fig. **1.17**). Indeed, a nonspecific opioid antagonist naltrexone (NTX) does not switch off the effect of ouabain–Ca^{2+} (Fig. **4.2.b**). These findings indicate that ouabain–Ca^{2+} can be compared to comenic acid or morphine in efficiency of $Na_V1.8$ channel modulation. It switches on the three background mechanisms discussed in Chapters **1-3** that should lead to pain relief: reduces the channels density, positively shifts $Na_V1.8$ channel activation gating process and, most importantly, markedly decreases Z_{eff}. The latter process is of dose-dependent nature, showing monotonic transducer-coupled ligand-receptor binding of ouabain–Ca^{2+} in subnanomolar and nanomolar concentration range from 100 pM to 10 nM (Fig. **4.3**). This binding process is characterized by $K_d = 7$ nM: extremely ouabain-sensitive branch of the dose-dependence curve reflects modulation of the signal-transducing function of Na^+, K^+-ATPase at the membrane level. As it was mentioned above, ouabain concentrations detected in human blood plasma are of the same order of magnitude (close to K_d). Thus, according to our results, neuronal Na^+,K^+-ATPase as a signal transducer should be under effective control of endogenous ouabain. On the contrary, the right branch of Z_{eff} dependence on ouabain concentration is governed by a radically different background mechanism. An increase of ouabain concentration leads to inhibition of the pumping function of Na^+,K^+-ATPase. The second process can also be approximated by the Hill equation, the K_d value in this case being much higher (0.1 mM) (Fig. **4.3**). This fact leads to raising a very important question. How can one and the same enzyme (Na^+,K^+-ATPase) distinguish between its pumping and non-pumping functions modulated by one and the same molecule (ouabain)? The answer is partly given above: endogenous ouabain exists in the form of calcium chelate complex due to the presence of free calcium in small amounts in physiological medium. An increase of ouabain concentration, given that calcium concentration remains the same, results in an important effect: "free" ouabain binds to a completely different site of Na^+,K^+-ATPase. This fact will be explained below on the basis of our quantum-chemical calculations. Here it is worth noting that experimental evaluation of ouabain K_d carried out by different methods never led to unequivocal results, which could be in part accounted for by the heterogeneity of Na^+,K^+-ATPase isoforms. Four isoforms of its α subunit are known to exist, and they are expressed in a cell type-specific manner in higher vertebrates. Adult rat kidney and liver cells express the α1 isoform; glial and skeletal muscle, both α1 and α2; sperm cells, both α1 and α4. Unlike most other cells, neurons may express α1, α2, α3, or any combination of these isoforms. With rare exceptions, the α3 isoform of Na^+,K^+-ATPase is detected in neurons of adult vertebrates only [20]. Therefore it is not surprising that quantitative data concerning the mechanisms of binding of cardiotonic steroids to Na^+,K^+-ATPase scatter significantly. The K_d values describing these processes vary with isoform type and they are different in rodents and humans. Human α1, α2, α3, and α4 isoforms have K_d in the range of 10^{-9} to 10^{-8} M [21 - 23]. One of these isoforms (α1) is ouabain-insensitive in rodents and has a very high K_d value of about 10^{-6} M [24, 25]. Furthermore, binding of cardiotonic steroids to the α subunit of Na^+,K^+-ATPase is affected by the particular β subunit associated with it [26]. Taking into account the three α subunits (α4 appears to be present specifically in spermatozoa), the three β subunits and the seven FXYD subunits that have been shown to associate with Na^+,K^+-ATPase, there are potentially 63 different receptor complexes with which cardiotonic steroids can interact [27].

Our patch-clamp data indicate that there are at least two different ouabain binding sites in Na^+,K^+-ATPase (Fig. **4.3**). Investigations of ouabain-sensitive current of Na^+,K^+-ATPase in small neurons from adult rat dorsal root ganglia also demonstrated the existence of two ouabain binding sites, which were suggested to be located on two functionally distinct Na^+,K^+-ATPase isozymes, α1β1 and α3β1, with ouabain dissociation constants of 0.2 and 140.1 μM, respectively [28].

(a)

1 nA

5 ms

(b)

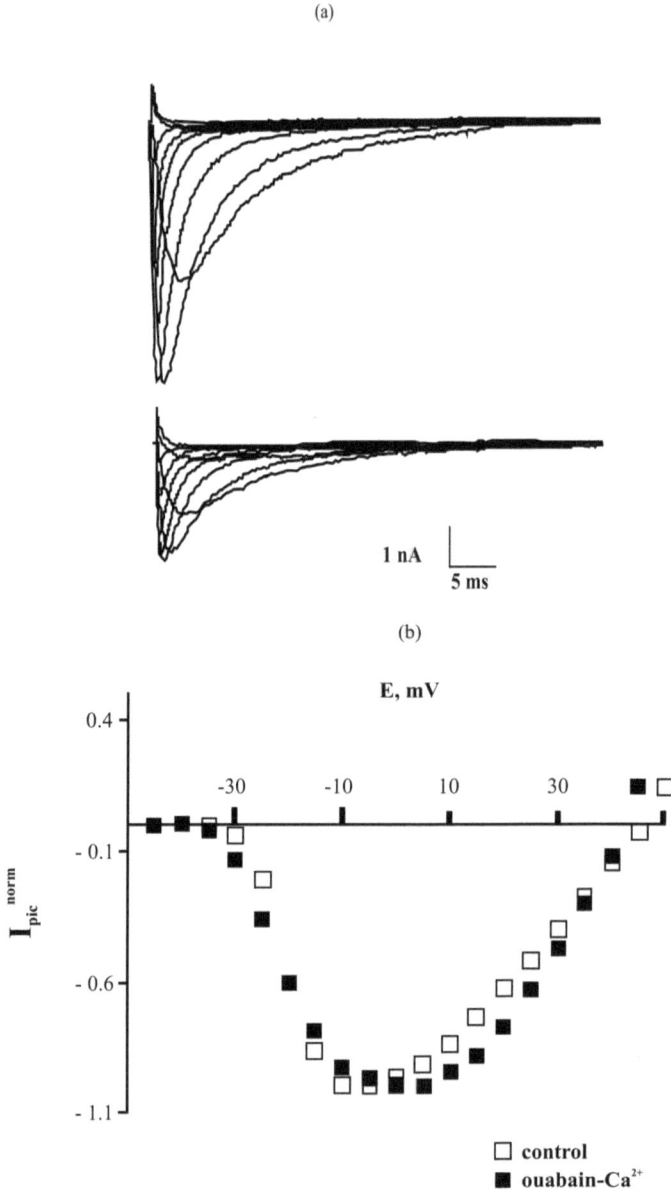

Fig. (4.1). Effects of ouabain on $Na_V1.8$ channels.
a – Families of sodium currents measured in the control experiment (top) and after application of ouabain at 1 nM (bottom); at this concentration the agent exists in the form of calcium chelate complex, ouabain–Ca^{2+};
b – Positive shift of the normalized peak current-voltage curve after application of ouabain–Ca^{2+}.

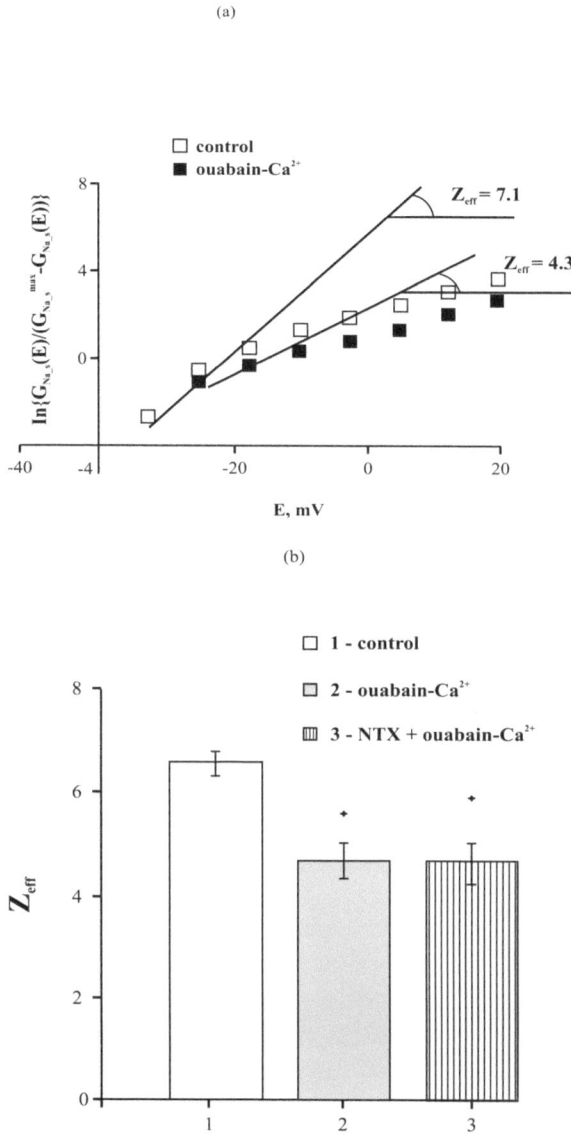

Fig. (4.2). Decrease of effective charge of $Na_V 1.8$ channels activation gating device after application of ouabain–Ca^{2+}:

a – Z_{eff} evaluation by the Almers' limiting slope procedure in the control experiment and after application of ouabain–Ca^{2+} at 1 nM;

b – ouabain–Ca^{2+} complex does not activate opioid-like receptors:

1 – the control value of Z_{eff} (6.5 ± 0.2, n = 15);

2 – Z_{eff} is reduced after application of ouabain–Ca^{2+} at 1 nM (4.8 ± 0.3, n = 15);

3 – combined application of NTX (50 μM) and ouabain-Ca^{2+} (1 nM) does not block the decrease of Z_{eff} (4.9 ± 0.3; n = 24);

* – difference between experimental and control data is statistically significant.

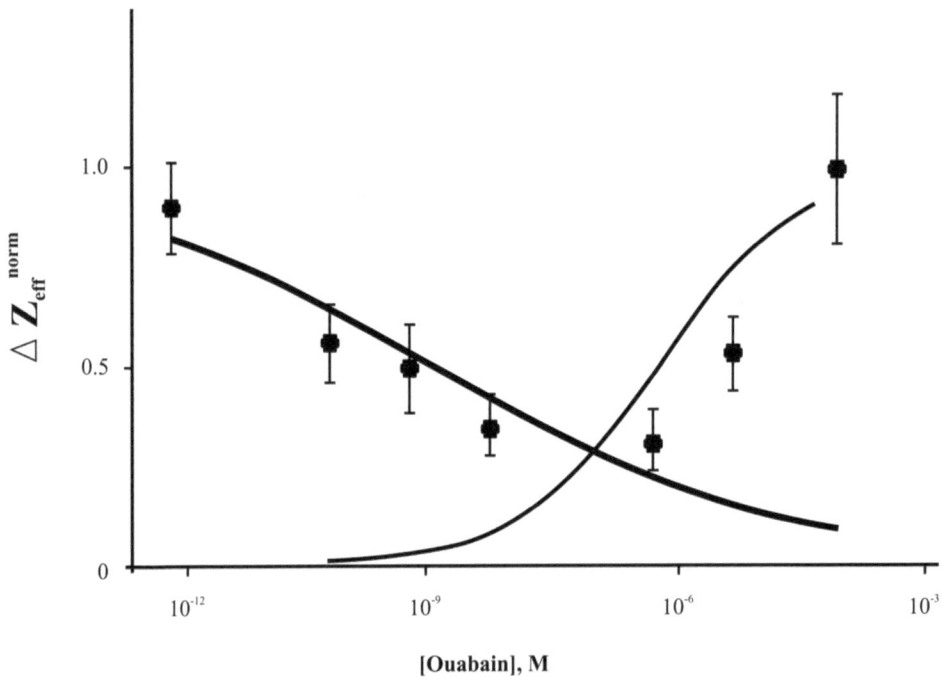

Fig. (4.3). Relationship between changes in the normalized effective charge ($\Delta Z_{\mathrm{eff}}^{\mathrm{norm}}$) and extracellular ouabain concentration.

$\Delta Z_{\mathrm{eff}}^{\mathrm{norm}} = (Z_{\mathrm{eff}} - Z_{\mathrm{eff}}^{\mathrm{min}})/(Z_{\mathrm{eff}}^{\mathrm{max}} - Z_{\mathrm{eff}}^{\mathrm{min}})$. The bold solid line displays the results obtained by application of the Hill equation ($K_d = 7$ nM, $n = 0.2$) to ouabain–Ca^{2+} binding at low concentrations. The thin solid line displays the results obtained by analogous calculations ($K_d = 0.1$ mM, $n = 0.5$) for the higher ouabain concentrations.

The fragment of ouabain molecule, ouabagenin, is structurally close to ouabain and also capable of chelating Ca^{2+}. However, our experiments show that it affects the functioning of $Na_V 1.8$ channels by another mechanism, which does not necessarily involve Na^+,K^+-ATPase. Application of ouabagenin (100 nM) evokes a decrease in the amplitude values of the currents and induces a positive voltage shift of the current-voltage characteristic (Fig. **4.4**). The value of Z_{eff} is also strongly decreased after application of ouabagenin, but its concentration was 100 times higher than that of ouabain (Fig. **4.5.a**). A very important our finding is that ouabagenin does not activate the signaling mechanism of Na^+,K^+-ATPase. This conclusion is based on the result that combined application of ouabain (200 μM) and ouabagenin (100 nM) does not block the effects of the latter (Fig. **4.5.b**), thus indicating that ouabagenin may be involved in the control of nociceptive signals but it does not activate the transducer-coupled mechanism in which Na^+,K^+-ATPase plays the central role.

(a)

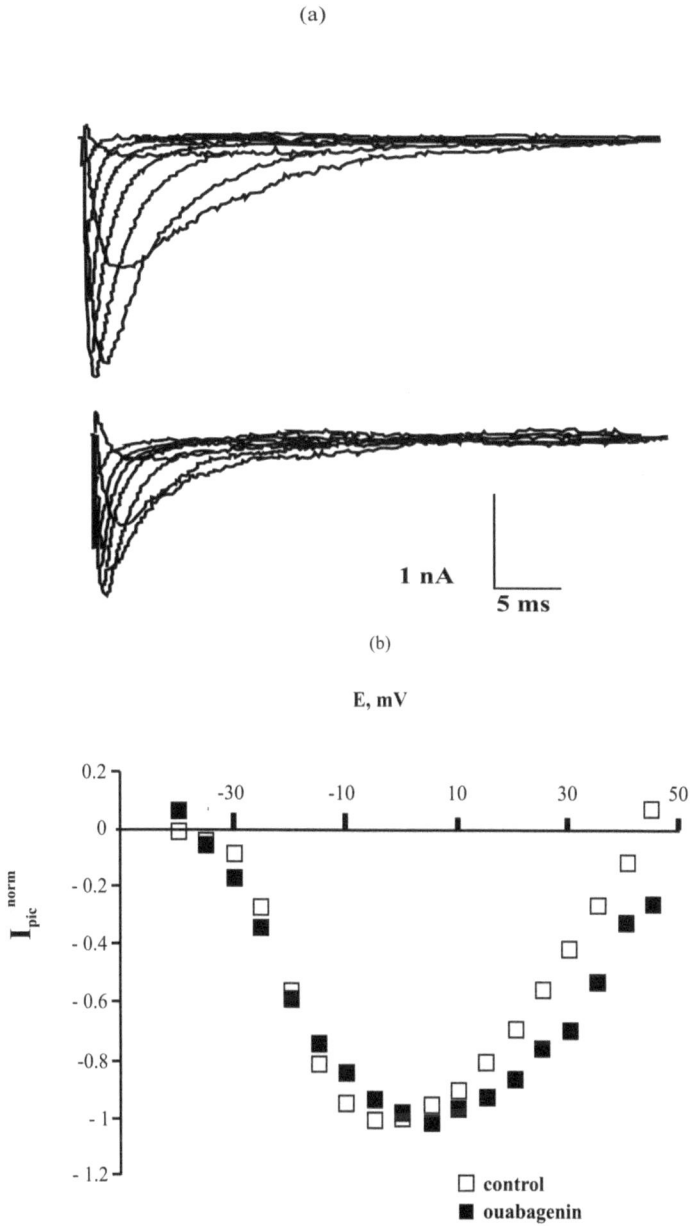

(b)

Fig. (4.4). Effects of ouabagenin on $Na_V 1.8$ channels.

a – Families of sodium currents measured in the control experiment (top) and after application of ouabagenin at 100 nM (bottom);
b – Positive shift of the normalized peak current-voltage curve after ouabagenin application.

(a)

(b)

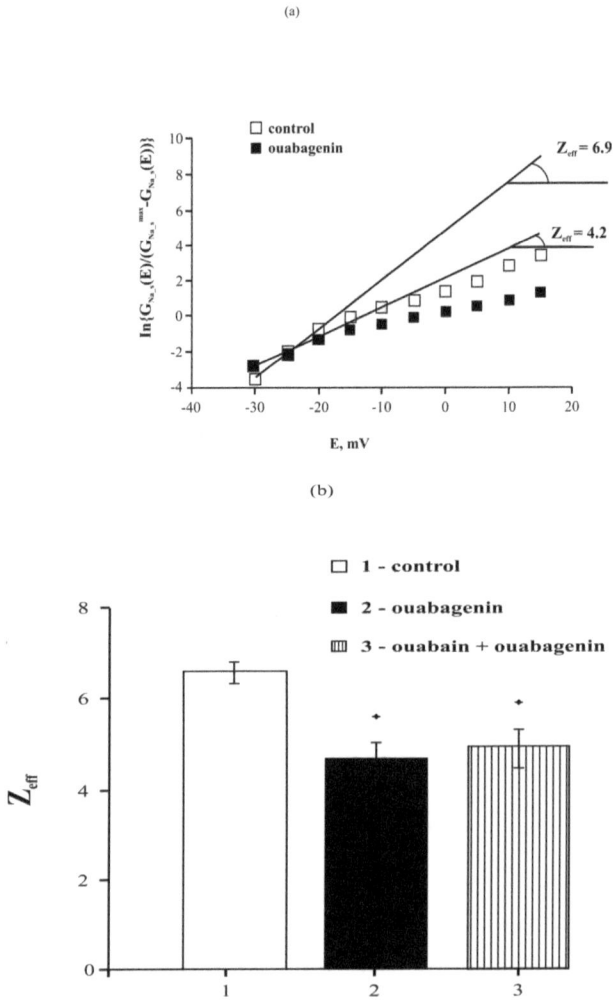

Fig. (4.5). Ouabagenin does not activate the transducer-coupled mechanism of modulation of $Na_V1.8$ channels.

a – The control value of Z_{eff} (white squares); Z_{eff} after ouabagenin (100 nM) application (black squares);

b – High ouabain concentration (200 μM) does not block the ouabagenin effect:

1 – the control value of Z_{eff} (6.7±0.6, n = 15);

2 – Z_{eff} (4.8±0.3, n = 9) after application of ouabagenin at 100 nM;

3 – combined application of ouabain (200 μM) and ouabagenin (100 nM) does not block the decrease of Z_{eff} (5.0±0.3, n = 20).

* – difference between experimental and control data is statistically significant

Somewhat unexpectedly, similar results were obtained examining another endogenous cardiotonic steroid, marinobufagenin. Its influence on the nociceptive system has not been investigated to date. Our experimental results demonstrate that this agent is also able to decrease $Na_V1.8$ currents (Fig. **4.6.a**) and induces a positive voltage shift of the current-voltage curve (Fig. **4.6.b**). The value of Z_{eff} is substantially reduced after marinobufagenin application (Fig. **4.7.a**). It should be stressed that Na^+,K^+-ATPase is not involved in marinobufagenin control of $Na_V1.8$ channels. Indeed, combined application of ouabain (200 μM) and marinobufagenin (10 nM) does not block the marinobufagenin effects (Fig. **4.7.b**). So, it can be concluded that this steroid interacts with the other target at extremely low endogenous concentrations. Our patch-clamp data indicate that both marinobufagenin and ouabagenin probably activate the modulated receptor mechanism through direct interaction with a specific binding site located on the channel itself.

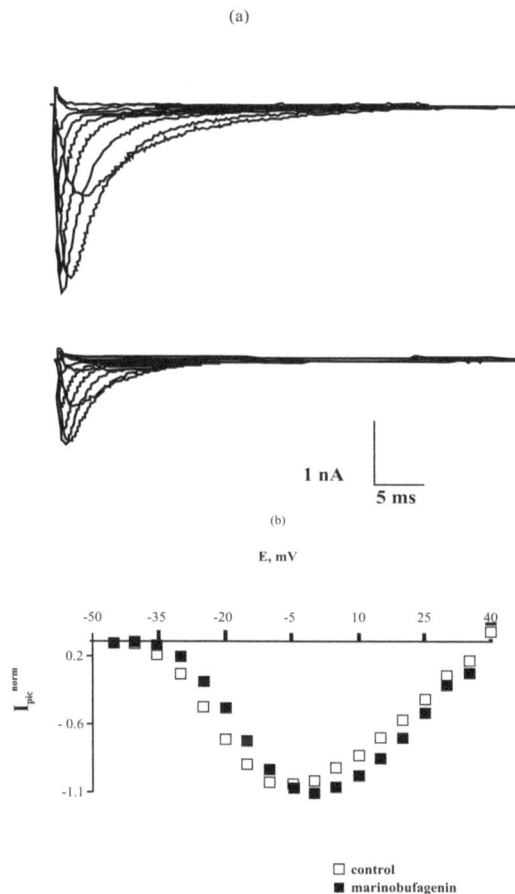

Fig. (4.6). Effects of marinobufagenin on $Na_V1.8$ channels.
a – Families of sodium currents measured in the control experiment (top) and after application of marinobufagenin at 10 nM (bottom);
b – Positive shift of the normalized peak current-voltage curve after marinobufagenin (10 nM) application.

(a)

(b)

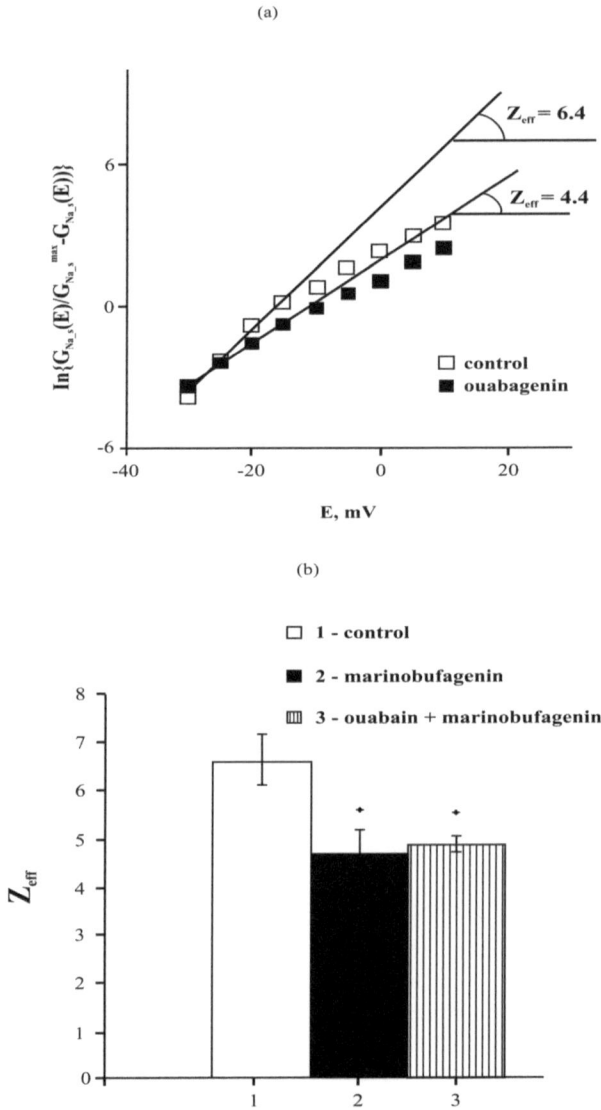

Fig. (4.7). Marinobufagenin does not activate the transducer-coupled mechanism of modulation of $Na_V1.8$ channels.

a – The control value of Z_{eff} (white squares); Z_{eff} after marinobufagenin (10 nM) application (black squares);

b – High ouabain concentration (200 μM) does not block the marinobufagenin effect:

1 – the control value of Z_{eff} (6.8± 0.4, n = 15);

2 – Z_{eff} (4.7±0.4, n = 15) after application of marinobufagenin at 10 nM;

3 – combined application of ouabain (200 μM) and marinobufagenin (10 nM) does not block the decrease of Z_{eff} (4.9±0.2, n = 9).

* – difference between experimental and control data is statistically significant.

Both morphine and comenic acid were shown above to control $Na_V 1.8$ channels by the opioid-like receptor-coupled mechanism (Figs. **1.16** and **2.3**). The current chapter presents the data indicating that endogenous ouabain (probably, in the form of ouabain–Ca^{2+} chelate complex) activates the transducing function of Na^+,K^+-ATPase. The effector unit of both mechanisms is $Na_V 1.8$ channel, which can also be under direct control of ouabagenin and marinobufagenin by the modulated receptor mechanism. In all these cases action of the agents is characterized by the same three main manifestations of inhibitory properties diminishing excitation of nociceptive neuron: they evoke a decrease of the amplitude values of $Na_V 1.8$ currents; they shift voltage sensitivity of $Na_V 1.8$ activation gating device in depolarizing direction and, finally, strongly decrease Z_{eff}. Therefore, during development of new analgesic drugs two additional targets for their active substances should be considered: opioid-like receptors and Na^+,K^+-ATPase, which plays the role of the signal transducer in nociceptive neuron membrane. Fig. (**1.17**) illustrates our main finding concerning this novel additional function of Na^+,K^+-ATPase: the enzyme was demonstrated to be included in the membrane signaling cascade (opioid-like receptor → Na^+,K^+-ATPase → $Na_V 1.8$ channel) as a serial unit [29]. This is the first indication on the alternative non-pumping function of Na^+,K^+-ATPase, which was published in Russian. English translation [30] of the original article was made without participation of the authors which led to mistakes in the text, as well as to probable misunderstanding of the obtained results by the readers. However, these findings allowed to explain the physiological function of extremely low endogenous concentrations of ouabain detected in human blood plasma [17 - 19]. Participation of Na^+,K^+-ATPase in processing of nociceptive information was never reported before. In this chapter we describe our novel results concerning the coupling between Na^+,K^+-ATPase and $Na_V 1.8$ channels. This transducer-coupled modulation of $Na_V 1.8$ channels seems to be a very promising mechanism for control of nociceptive signals. Involvement of Na^+,K^+-ATPase in this mechanism makes the enzyme a target for novel analgesic drugs. Also, in terms of fundamental physiology, it is equally important to understand mechanisms of action of endogenous substances, such as ouabain and marinobufagenin, which control the functioning of the nociceptive system at the molecular level.

At the same time the non-pumping function of Na^+,K^+-ATPase was discovered in cardiac myocytes [31]. The Na^+,K^+-ATPase signal-transducing function appears to have been acquired through the evolutionary incorporation of many specific binding motifs designed to interact with various other ligands and proteins. Binding of ouabain to Na^+,K^+-ATPase activates different signaling modules. On the one hand, the crosstalk between the activated pathways eventually modulates the expression of a number of genes [31]. That is why ouabain influences the growth of cardiac myocytes and the nerve tissue as well as proliferation of the smooth muscle cells; it also induces apoptosis in various malignant cells [32].

Nanomolar concentrations of cardiotonic steroids, such as ouabain, digoxin, marinobufagenin, bufalin and telocinobufagin, detected in blood plasma of experimental animals and humans were shown to be in the range of 0.5–20 nM [17 - 19, 33 - 43]. These values were considered too low to modulate the Na^+,K^+-ATPase pumping activity and thus were not expected to affect intracellular Na^+ and Ca^{2+} concentrations [44]. Therefore, a hypothesis discussed above that Na^+,K^+-ATPase can also act as a receptor for cardiotonic steroids has been presented: indeed, evidence is accumulating that, in addition to its transport function, Na^+,K^+-ATPase also plays a signaling role, regulating early response genes associated with cell growth [32, 45].

Though the mechanisms of ouabain control of the Na^+,K^+-ATPase pumping function are well studied [46], the background mechanisms of the mitogenic effect of ouabain are not completely understood. We suggest that this effect is accounted for by activation of the signal-transducing function of Na^+,K^+-ATPase [47, 48]. Quantum-chemical calculations should be of much help in elucidating the ability of ouabain to selectively modulate both pumping and non-pumping functions of Na^+,K^+-ATPase. Assuming that these two functions are controlled by different molecular forms of ouabain (*e.g.*, conformers or complexes of ouabain with inorganic cations), an existence of two ouabain binding sites on the same Na^+,K^+-ATPase molecule might be considered: one of them regulates the pumping function of the enzyme, while the other the signal-transducing function.

A calculational study on all conformers of the examined steroids and their ability to chelate Ca^{2+} can provide an answer to both questions, making it possible to find two distinct molecular forms of ouabain which account for its ability to control both pumping and signal-transducing functions of Na^+,K^+-ATPase and to reveal structural features of ouabain, ouabagenin, and marinobufagenin which explain the fact that these agents modulate $Na_V1.8$ channels by different molecular mechanisms.

QUANTUM-CHEMICAL STUDY OF CARDIOTONIC STEROIDS

Ouabain and Ouabagenin

Despite a significant physiological importance of both molecules, only a few studies were dedicated to investigation of their structure. The stereochemical formulae of ouabain and ouabagenin are shown in Fig. (**4.8**). X-ray analysis has been performed in the early 1980s for ouabain crystal hydrate [49], complex of ouabain with ethanol [50], and complex of ouabagenin with methanol [51]. Single crystal X-ray diffraction analysis seldom provides information regarding the conformational space of examined molecules. Steric structures of ouabain and ouabagenin were also studied by NMR spectroscopy with subsequent molecular dynamics resolution [52]. Both molecules were shown to exhibit a considerable conformational rigidity due to the presence of four fused rings. Only ring A of the steroid core and the lactone ring have certain mobility.

(b)

Fig. (4.8). Stereochemical formulae of ouabain (a) and ouabagenin (b).

The electronic structure of ouabain and ouabagenin is poorly studied. Several decades ago, the electrostatic potentials of a series of cardiotonic steroids were calculated by the semiempirical CNDO method [53]. However, the computer performance was insufficient to apply *ab initio* calculational methods to these relatively large molecules. Since that time, the interest in quantum-chemical studies of cardiotonic steroids has noticeably decreased, while examination of the electronic structure of these physiologically important molecules on a high level of theory could give an insight into the mechanism of their physiological action.

The geometric parameters of isolated ouabain and ouabagenin molecules were fully optimized *ab initio* by the restricted Hartree-Fock method using the 6-31G* basis set [54] and GAMESS program [55]. Four possible conformations were considered for each of the molecules. Rings B and C of the steroid core were demonstrated previously to adopt the *chair* conformation only; ring D, the 14β-*envelope* conformation; and ring A, either the *chair* (more energetically favorable) or the *twist* conformations [52, 56]. The lactone ring can have two orientations relative to the remaining part of the molecule. Rotation around the C—C bond linking the lactone ring to the steroid core is rather unrestricted, and two energy minima appear on the potential energy surface [49 - 51, 57].

Steric structures of selected ouabain and ouabagenin conformations, the atom numbering, and the patterns of intramolecular hydrogen bonding are shown in Fig. (**4.9**). Ouabain and ouabagenin conformations are denoted as follows: indices C and T refer to the conformation of ring A (*chair* and *twist*, respectively); indices 1 and 2, to the orientation of the lactone group relative to the steroid core ($C^{16}C^{17}C^{20}C^{21}$ torsion angle values about 135° and -52°, respectively). The ouabagenin molecule differs from the ouabain molecule in the O^3 atom substituent (H^3 atom or rhamnosyl residue, respectively).

Our calculations indicate that C2 conformations of both molecules have the lowest energy in the gas phase. The energies of C1 conformations are higher by ~1 kcal·mol^{-1}; the energies of T2 conformations, by ~8 and ~7 kcal·mol^{-1} for ouabain and ouabagenin, respectively; the energies of T1 conformations are the highest (~9 and ~8 kcal·mol^{-1}, respectively, relative to C2). Ouabain was demonstrated by X-ray diffraction analysis to crystallize from water in C1 conformation [49] and from ethanol in C2 conformation [50]. Ouabagenin crystallizes from methanol in C2 conformation [51]. A liquid-phase NMR study showed that both molecules adopted the lowest-energy C1 conformations in D_2O and DMSO-d_6/CDCl$_3$, with interconversion to T1 being possible [52]. An increase in the dielectric constant of the solvent likely results in a decrease of the energy of C1 conformation relative to C2.

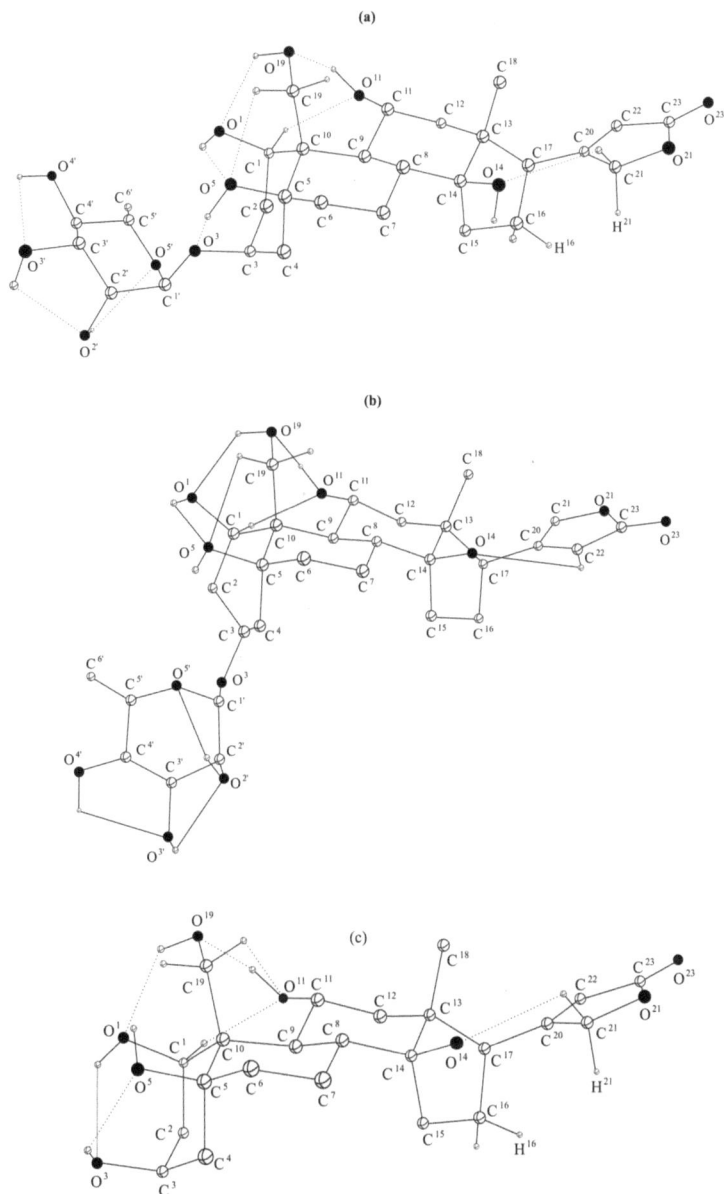

Fig. (4.9). Steric structures of selected ouabain and ouabagenin conformations, accepted atom numbering, and intramolecular hydrogen bonding patterns:
a – Conformation C2 of ouabain;
b – Conformation T1 of ouabain;
c – Conformation C2 of ouabagenin.
Only some of hydrogen atoms are shown. Hydrogen bonds are presented as dotted lines.

Selected bond lengths, bond angles, torsion angles, and atomic charges in all examined conformations of ouabain and ouabagenin are presented in Table **4.1**. The results of single crystal X-ray diffraction studies of ouabain [50] and ouabagenin [51] in C2 conformation are also given for comparison. For ouabagenin, the agreement between the calculated and experimentally obtained geometric parameters is reasonably good, whereas it is slightly worse for ouabain. This primarily concerns C^1—O^1 and C^{21}—O^{21} bond lengths: the values calculated *ab initio* are 0.21-0.46 Å shorter than the experimental values. Certain differences are observed between the optimized and experimentally determined C—C bond lengths in the steroid core. The lactone ring is calculated to be planar (values of torsion angles within the ring $|\tau|$ do not exceed 0.5°), while it is distorted in the experiment ($|\tau| \leq 5°$) [49]. The steroid core rings are also somewhat more planar: the simple arithmetic average of the absolute values of endocyclic torsion angles (τ_{av}) for rings A, B, C, and the rhamnose ring in C1 conformation of ouabain are 51.9°, 51.4°, 55.7°, and 55.3°, respectively, according to the calculations, and 50.0°, 61.5°, 58.8°, and 58.0°, according to single crystal X-ray diffraction results [49]. The differences between the optimized and experimental geometry parameters of ouabain and ouabagenin are accounted for by the fact that the effects of the solvent medium and molecular packing in the crystal are not considered in our calculations. Moreover, the molecules in the gas phase are not involved in intermolecular hydrogen bonding detected in the crystal lattice [49 - 51]. Nevertheless, the structural parameters of ouabain and ouabagenin conformers obtained by RHF/6-31G* *ab initio* calculations quite satisfactorily agree with the experimental data: the low-energy conformations are reproduced adequately, and the accordance of the bond lengths, bond and torsion angles with the experiment is acceptable.

Table 4.1. Selected calculated (RHF/6-31G*) and experimental bond lengths, bond and torsion angles, and atomic charges in ouabain and ouabagenin.

Parameter	Ouabain					Ouabagenin				
	C1	C2	T1	T2	Experiment	C1	C2	T1	T2	Experiment
Bond Length, r, Å										
C^1—C^2	1.531	1.531	1.541	1.542	1.532	1.529	1.529	1.540	1.540	1.531
C^1—C^{10}	1.568	1.568	1.582	1.582	1.578	1.570	1.571	1.582	1.582	1.564
C^2—C^3	1.522	1.522	1.513	1.513	1.532	1.520	1.519	1.513	1.513	1.517
C^4—C^5	1.541	1.541	1.553	1.553	1.523	1.538	1.539	1.553	1.553	1.543
C^5—C^{10}	1.568	1.568	1.556	1.556	1.578	1.576	1.577	1.556	1.557	1.571
C^{13}—C^{14}	1.550	1.546	1.551	1.546	1.594	1.549	1.545	1.551	1.547	1.540

(Table 4.1) contd.....

Parameter	Ouabain					Ouabagenin				
	C1	C2	T1	T2	Experiment	C1	C2	T1	T2	Experiment
C^{13}—C^{17}	1.582	1.572	1.583	1.572	1.550	1.582	1.572	1.582	1.572	1.588
C^{15}—C^{16}	1.531	1.537	1.530	1.537	1.537	1.531	1.537	1.531	1.537	1.534
C^{16}—C^{17}	1.553	1.557	1.552	1.557	1.523	1.552	1.557	1.552	1.557	1.568
C^{20}—C^{21}	1.508	1.514	1.508	1.514	1.489	1.508	1.514	1.508	1.514	1.494
C^{1}—O^{1}	1.414	1.414	1.417	1.417	1.460	1.413	1.413	1.418	1.418	1.448
C^{3}—O^{3}	1.424	1.424	1.416	1.416	1.416	1.414	1.413	1.408	1.408	1.425
C^{21}—O^{21}	1.405	1.412	1.405	1.412	1.435	1.405	1.411	1.405	1.412	
C^{23}—O^{21}	1.346	1.342	1.346	1.342	1.335	1.346	1.342	1.346	1.342	
Bond Angle, ω, deg										
$C^{1}C^{2}C^{3}$	115.4	115.3	113.2	113.4	114.3	113.4	113.4	114.0	114.0	114.9
$C^{1}C^{10}C^{9}$	111.9	111.9	113.8	113.9		110.3	110.3	113.8	113.7	111.8
$C^{2}C^{1}C^{10}$	115.1	115.1	116.5	116.6	113.0	115.2	115.2	116.1	116.1	113.0
$C^{3}C^{4}C^{5}$	114.7	114.7	116.0	116.1	114.9	114.3	114.3	116.0	116.1	115.5
$C^{4}C^{5}C^{10}$	112.1	112.1	113.1	113.1	112.7	112.3	112.3	113.1	113.3	111.7
$C^{5}C^{10}C^{1}$	107.6	107.6	106.8	106.8	107.4	109.1	109.1	107.0	107.0	108.0
$C^{5}C^{10}C^{9}$	110.3	110.2	108.8	108.8	109.5	110.4	110.4	108.8	108.9	108.3
$C^{5}C^{10}C^{19}$	109.5	109.6	110.6	110.6		109.5	109.6	110.6	110.6	109.9
$C^{6}C^{7}C^{8}$	111.7	111.6	113.1	112.9	111.7	111.3	111.3	113.1	113.0	111.9
$C^{7}C^{8}C^{9}$	114.7	114.7	115.7	115.5	115.3	114.3	114.3	115.8	115.6	111.4
$C^{8}C^{9}C^{10}$	113.8	113.7	113.0	112.9	113.3	114.5	114.4	113.1	113.0	110.7
$C^{9}C^{10}C^{19}$	109.9	109.9	108.9	108.9	111.5	110.0	110.0	109.0	109.0	110.5
$C^{14}C^{13}C^{18}$	112.8	114.0	112.7	114.0		112.8	114.0	112.8	114.0	114.1
$C^{14}C^{15}C^{16}$	103.7	104.9	103.6	104.9	104.6	103.7	105.0	103.7	105.0	104.5
$C^{15}C^{16}C^{17}$	106.0	106.8	106.0	106.8	107.1	106.1	106.8	106.0	106.8	106.3
$C^{17}C^{20}C^{21}$	121.0	126.6	121.1	126.6	125.0	121.0	126.6	121.0	126.6	126.3
$C^{17}C^{20}C^{22}$	131.3	126.0	131.3	126.0	125.7	131.3	125.9	131.3	126.0	125.8
$C^{2}C^{1}O^{1}$	110.9	111.0	107.4	107.0	109.5	110.7	110.7	107.3	107.3	110.1
$C^{2}C^{3}O^{3}$	113.4	113.3	113.2	113.0	107.3	109.4	109.4	111.9	111.9	108.3
$C^{4}C^{3}O^{3}$	108.8	108.8	106.1	106.0	113.6	113.2	113.2	106.7	106.6	112.8
$C^{4}C^{5}O^{5}$	108.5	108.5	108.5	108.5	108.2	103.4	103.4	108.5	108.5	107.4
$C^{6}C^{5}O^{5}$	107.6	107.6	109.3	109.3	105.2	107.7	107.7	109.2	109.3	108.3
$C^{10}C^{5}O^{5}$	107.0	107.0	104.5	104.5		111.2	111.1	104.5	104.4	107.5
$C^{3}O^{3}C^{1'}$	117.4	117.2	118.3	118.5	115.1					

(Table 4.1) contd.....

Parameter	Ouabain					Ouabagenin				
	C1	C2	T1	T2	Experiment	C1	C2	T1	T2	Experiment
Torsion Angle, τ, deg										
$C^1C^2C^3C^4$	-48.4	-48.5	58.5	58.0		-54.1	-54.1	57.7	57.9	
$C^2C^3C^4C^5$	52.7	52.5	-29.6	-29.1		56.6	56.6	-28.8	-29.1	
$C^3C^4C^5C^{10}$	-57.6	-57.4	-29.4	-29.6		-55.5	-55.5	-29.7	-29.2	
$C^4C^5C^{10}C^1$	52.7	52.6	58.2	58.1		48.2	48.1	58.4	58.0	
$C^5C^{10}C^1C^2$	-49.9	-50.0	-29.8	-29.7		-47.8	-47.8	-30.1	-29.8	
$C^{10}C^1C^2C^3$	50.1	50.3	-27.6	-27.3		52.6	52.6	-26.9	-27.1	
$C^5C^6C^7C^8$	55.7	55.8	50.5	50.9		57.1	56.4	50.4	50.8	
$C^6C^7C^8C^9$	-48.4	-48.5	-43.2	-44.0		-49.9	-50.0	-42.8	-43.6	
$C^7C^8C^9C^{10}$	44.9	45.1	43.9	44.6		45.1	45.3	43.6	44.2	
$C^8C^9C^{10}C^5$	-46.5	-46.6	-50.3	-50.4		-44.6	-44.7	-50.3	-50.2	
$C^9C^{10}C^5C^6$	53.4	53.9	59.0	58.7		51.0	51.0	59.3	58.8	
$C^{10}C^5C^6C^7$	-59.7	-59.6	-60.1	-60.0		-58.7	-58.7	-60.3	-60.1	
$C^8C^9C^{11}C^{12}$	61.2	60.5	62.2	61.0		60.8	60.1	62.3	61.3	
$C^9C^{11}C^{12}C^{13}$	-61.0	-60.7	-61.6	-61.2		-61.0	-60.7	-61.5	-61.2	
$C^{11}C^{12}C^{13}C^{14}$	50.9	51.9	50.9	52.1		51.1	51.9	50.9	51.9	
$C^{12}C^{13}C^{14}C^8$	-46.9	-48.7	-46.6	-49.0		-47.0	-48.7	-46.9	-48.8	
$C^{13}C^{14}C^8C^9$	54.9	56.2	54.9	56.5		55.0	56.3	55.3	56.5	
$C^{14}C^8C^9C^{11}$	-59.5	-59.3	-60.2	-59.7		-59.3	-59.1	-60.5	-60.0	
$C^{13}C^{14}C^{15}C^{16}$	41.9	37.7	42.1	37.9		41.8	37.3	41.9	37.5	
$C^{14}C^{15}C^{16}C^{17}$	-32.5	-23.8	-33.2	-24.1		-32.4	-23.3	-32.5	-23.5	
$C^{15}C^{16}C^{17}C^{13}$	10.8	0.9	11.7	1.1		10.6	0.5	10.8	0.6	
$C^{16}C^{17}C^{13}C^{14}$	14.8	22.0	14.0	22.0		14.9	22.3	14.7	22.2	
$C^{17}C^{13}C^{14}C^{15}$	-34.7	-36.7	-34.3	-36.9		-34.7	-36.7	-34.7	-36.8	
$C^4C^5C^{10}C^9$	-69.7	-69.7	-65.1	-65.3		-73.2	-73.2	-64.8	-65.3	
$C^7C^8C^9C^{11}$	171.9	172.3	171.1	171.7		172.5	172.8	170.8	171.5	
$C^{12}C^{13}C^{14}C^{15}$	80.1	77.8	80.6	77.8		80.0	77.8	80.2	77.9	
$C^{16}C^{17}C^{20}C^{21}$	134.1	-52.6	137.0	-52.7		134.9	-52.9	134.5	-52.0	
$C^{16}C^{17}C^{20}C^{22}$	-43.9	124.0	-41.1	124.0		-43.3	123.7	-43.8	124.9	
Atomic Charge, q, a.u.										
C^2	-0.35	-0.35	-0.37	-0.37		-0.36	-0.36	-0.36	-0.36	
C^3	0.16	0.16	0.19	0.19		0.18	0.18	0.18	0.18	
C^5	0.32	0.32	0.31	0.31		0.31	0.31	0.31	0.31	

(Table 4.1) contd.....

Parameter	Ouabain					Ouabagenin				
	C1	C2	T1	T2	Experiment	C1	C2	T1	T2	Experiment
C^6	-0.31	-0.31	-0.33	-0.33		-0.34	-0.34	-0.33	-0.33	
C^{10}	-0.11	-0.11	-0.11	-0.11		-0.09	-0.09	-0.11	-0.11	
C^{19}	0.00	0.00	0.01	0.01		0.02	0.02	0.01	0.01	
O^3	-0.71	-0.71	-0.69	-0.69		-0.79	-0.79	-0.76	-0.76	

The calculated geometry parameters and Mulliken atomic charges that do not differ substantially in all examined ouabain and ouabagenin conformations are presented in Table **4.2** using the data obtained for C2 conformation of ouabain. The geometry parameters of rings B and C, the lactone and rhamnosyl rings are essentially independent on the overall conformation of the molecules. We have also performed full RHF/6-31G* geometry optimization of isolated rhamnose and lactone molecules and determined that the steric and electronic structure of these functional groups is fairly the same when they are considered as separate molecules and as fragments of ouabain and ouabagenin: the bond lengths change by less than 0.01 Å; the bond and torsion angles, by less than 1°. Noticeable differences in the atomic charges q are observed for C^{20} and C^{22} atoms only: $q(C^{20})$ = -0.21 a.u., $q(C^{22})$ = -0.29 a.u. in isolated lactone; $q(C^{20})$ = 0.03 a.u., $q(C^{22})$ = -0.36 a.u. in ouabain, which indicates that the lactone substituent is capable of donating the π-electron density of C^{20}—C^{22} double bond. However, conjugation of C^{20}—C^{22} and C^{23}—O^{23} double bonds is weakly pronounced in the lactone molecule, so this effect is not manifested in the remaining part of the lactone ring. Thus, the lactone and rhamnosyl rings can be regarded as independent fragments, the electronic and steric structure of which is not appreciably influenced by the steroid core. On the other hand, the steric orientation of these substituents affects the geometry of the steroid core and electron density distribution in it.

Table 4.2. Selected calculated (RHF/6-31G*) bond lengths, bond angles, and atomic charges in conformation C2 of ouabain.

Bond Length, r, Å									
C^3—C^4	1.523	C^9—C^{11}	1.551	C^{20}—C^{22}	1.323	$C^{1'}$—$C^{2'}$	1.525	$C^{2'}$—$O^{2'}$	1.409
C^5—C^6	1.526	C^{10}—C^{19}	1.556	C^{22}—C^{23}	1.477	$C^{2'}$—$C^{3'}$	1.525	$C^{3'}$—$O^{3'}$	1.402
C^6—C^7	1.523	C^{11}—C^{12}	1.527	C^5—O^5	1.425	$C^{3'}$—$C^{4'}$	1.525	$C^{4'}$—$O^{4'}$	1.396
C^7—C^8	1.535	C^{12}—C^{13}	1.542	C^{11}—O^{11}	1.401	$C^{4'}$—$C^{5'}$	1.526	$C^{5'}$—$O^{5'}$	1.420
C^8—C^9	1.563	C^{13}—C^{18}	1.536	C^{14}—O^{14}	1.418	$C^{5'}$—$C^{6'}$	1.516		
C^8—C^{14}	1.550	C^{14}—C^{15}	1.538	C^{19}—O^{19}	1.409	$C^{1'}$—O^3	1.385		
C^9—C^{10}	1.579	C^{17}—C^{20}	1.510	C^{23}—O^{23}	1.183	$C^{1'}$—$O^{5'}$	1.389		

(Table 4.2) contd.....

Bond Angle, ω, deg									
$C^1C^{10}C^{19}$	107.5	$C^{10}C^9C^{11}$	115.3	$C^{20}C^{22}C^{23}$	109.5	$C^{22}C^{23}O^{21}$	107.3	$C^{2'}C^{3'}O^{3'}$	111.3
$C^2C^3C^4$	109.6	$C^{11}C^{12}C^{13}$	114.8	$C^{21}C^{20}C^{22}$	107.4	$C^{22}C^{23}O^{23}$	129.4	$C^{3'}C^{2'}O^{2'}$	110.0
$C^4C^5C^6$	109.4	$C^{12}C^{13}C^{14}$	108.3	$C^8C^{14}O^{14}$	109.6	$O^{21}C^{23}O^{23}$	123.2	$C^{3'}C^{4'}O^{4'}$	110.7
$C^5C^6C^7$	112.3	$C^{12}C^{13}C^{17}$	107.8	$C^9C^{11}O^{11}$	114.8	$C^{1'}C^{2'}C^{3'}$	110.8	$C^{4'}C^{3'}O^{3'}$	110.5
$C^6C^5C^{10}$	111.8	$C^{12}C^{13}C^{18}$	109.0	$C^{10}C^1O^1$	110.7	$C^{2'}C^{3'}C^{4'}$	110.6	$C^{4'}C^{5'}O^{5'}$	108.9
$C^7C^8C^{14}$	111.2	$C^{13}C^{14}C^{15}$	103.6	$C^{10}C^{19}O^{19}$	112.7	$C^{3'}C^{4'}C^{5'}$	110.4	$C^{5'}C^{4'}O^{4'}$	108.4
$C^8C^9C^{11}$	105.3	$C^{13}C^{17}C^{16}$	105.6	$C^{12}C^{11}O^{11}$	105.9	$C^{4'}C^{5'}C^{6'}$	113.1	$C^{6'}C^{5'}O^{5'}$	107.0
$C^8C^{14}C^{13}$	113.5	$C^{13}C^{17}C^{20}$	117.4	$C^{13}C^{14}O^{14}$	105.6	$C^{1'}C^{2'}O^{2'}$	109.7	$O^{3'}C^{1'}O^{5'}$	113.6
$C^8C^{14}C^{15}$	115.9	$C^{14}C^{13}C^{17}$	104.1	$C^{15}C^{14}O^{14}$	108.0	$C^{1'}O^{5'}C^{5'}$	116.8		
$C^9C^8C^{14}$	112.6	$C^{16}C^{17}C^{20}$	112.3	$C^{20}C^{21}O^{21}$	104.8	$C^{2'}C^{1'}O^3$	108.6		
$C^9C^{11}C^{12}$	110.6	$C^{17}C^{13}C^{19}$	113.2	$C^{21}O^{21}C^{23}$	111.0	$C^{2'}C^{1'}O^{5'}$	110.2		
Atomic Charge, q, a.u.									
C^1	0.16	C^{10}	-0.11	C^{18}	-0.48	O^5	-0.81	$C^{3'}$	0.14
C^2	-0.35	C^{11}	0.21	C^{19}	0.00	O^{11}	-0.79	$C^{4'}$	0.16
C^3	0.16	C^{12}	-0.33	C^{20}	0.03	O^{14}	-0.79	$C^{5'}$	0.12
C^4	-0.37	C^{13}	-0.05	C^{21}	0.01	O^{19}	-0.79	$C^{6'}$	-0.49
C^5	0.39	C^{14}	0.34	C^{22}	-0.36	O^{21}	-0.63	$O^{2'}$	-0.78
C^6	-0.31	C^{15}	-0.35	C^{23}	0.81	O^{23}	-0.57	$O^{3'}$	-0.78
C^7	-0.36	C^{16}	-0.33	O^1	-0.81	$C^{1'}$	0.45	$O^{4'}$	-0.77
C^8	-0.18	C^{17}	-0.20	O^3	-0.71	$C^{2'}$	0.11	$O^{5'}$	-0.69
C^9	-0.22								

The results of quantum-chemical calculations reveal the effect of three major factors on the structure of the steroid core of ouabain and ouabagenin: 1) orientation of the lactone ring relative to the steroid core (comparison of the parameters of conformation families 1 and 2); 2) ring A conformation (comparison of the parameters of conformation families C and T); 3) presence of the rhamnosyl residue (comparison of the parameters of ouabain and ouabagenin molecules in respective conformations). A conformation family is considered to include all conformations sharing a certain structural feature, *e.g.*, family C includes all conformations of ouabain and ouabagenin in which ring A adopts the *chair* conformation.

The differences in geometry between conformation families 1 and 2 are largely determined by intramolecular hydrogen bonding patterns. Weak O···H—C hydrogen bonds were detected in both conformation families. O^{14}···H—C^{22} bond in conformation family 1 ($r(O^{14}···C^{22})$ distance ≈ 3.45 Å) is noticeably weaker than

$O^{14}\cdots H - C^{21}$ bond in conformation family 2 ($r(O^{14}\cdots C^{21}) \approx 3.10$ Å).

Furthermore, conformations of family 2 are additionally stabilized by attraction of two hydrogen atoms, H^{16} and H^{21} (Figs. **4.9.a** and **4.9.c**), which influences the geometry of ring D. Ring D in conformation family 1 has a somewhat distorted shape ($\tau(C^{15}C^{16}C^{17}C^{13}) \approx 11°$), whereas in conformation family 2 it adopts the undistorted 14β-*envelope* conformation ($\tau(C^{15}C^{16}C^{17}C^{13}) \approx 0°$). Endocyclic torsion angles in ring D and torsion angles defining mutual orientation of rings C, D, and the lactone ring in examined conformations are given in Table **4.1**. The geometry of the lactone ring remains essentially unchanged. Rotation of the lactone moiety around $C^{17} - C^{20}$ bond affects the values of $C^{17}C^{20}O^{21}$ and $C^{17}C^{20}O^{22}$ bond angles: they are close in conformation family 2, while differing by ~10° in conformation family 1. Electron density redistribution in ouabain and ouabagenin molecules associated with lactone rotation is insignificant.

The effect of the rhamnosyl substituent and ring A conformation on structural and electronic parameters of ouabain and ouabagenin can be clearly revealed by analyzing intramolecular hydrogen bonds detected in studied conformers. The hydrogen bond parameters are given in Table **4.3**. When ring A adopts the *twist* conformation, O^3 atom becomes too remote from other oxygen atoms and incapable of participating in hydrogen bonding. Therefore, all conformations T of ouabain and ouabagenin share a similar pattern of hydrogen bonds. If ring A is in the *chair* conformation, certain differences in intramolecular hydrogen bonding patterns are observed between ouabain and ouabagenin. In the ouabain molecule, O^3 atom is linked to the rhamnosyl residue, thus being able only to accept a proton, whereas in the ouabagenin molecule this atom is a part of the hydroxyl group which can also act as the proton donor.

Table 4.3. Calculated parameters of intramolecular hydrogen bonds[a] present in examined conformations of ouabain and ouabagenin: R_1 (Å), O⋯H distance; R_2 (Å), O⋯X distance.

Ouabain											
C1			C2			T1			T2		
O⋯X atoms	R_1	R_2	O⋯X atoms	R_1	R_2	O⋯X atoms	R_1	R_2	O⋯X atoms	R_1	R_2
$O^5\cdots O^1$	2.01	2.77	$O^5\cdots O^1$	2.01	2.77	$O^1\cdots O^{19}$	2.00	2.66	$O^1\cdots O^{19}$	2.00	2.66
$O^{19}\cdots O^{11}$	2.08	2.86	$O^{19}\cdots O^{11}$	2.10	2.87	$O^{19}\cdots O^{11}$	2.13	2.88	$O^{19}\cdots O^{11}$	2.16	2.90
$O^1\cdots O^{19}$	2.15	2.77	$O^1\cdots O^{19}$	2.15	2.77	$O^{11}\cdots C^1$	2.21	3.04	$O^{11}\cdots C^1$	2.21	3.04
$O^{11}\cdots C^1$	2.21	3.02	$O^{11}\cdots C^1$	2.20	3.02	$O^5\cdots O^1$	2.26	2.90	$O^5\cdots O^1$	2.27	2.90
$O^3\cdots O^5$	2.26	2.77	$O^3\cdots O^5$	2.25	2.77	$O^2{}'\cdots O^3{}'$	2.30	2.79	$O^2{}'\cdots O^3{}'$	2.30	2.78
$O^5\cdots C^{19}$	2.27	2.78	$O^5\cdots C^{19}$	2.27	2.78	$O^5\cdots C^{19}$	2.36	2.81	$O^5\cdots C^{19}$	2.35	2.81
$O^2{}'\cdots O^3{}'$	2.29	2.78	$O^2{}'\cdots O^3{}'$	2.29	2.78	$O^5\cdots O^2{}'$	2.42	2.86	$O^5\cdots O^2{}'$	2.42	2.86

(Table 4.3) contd.....

Ouabain											
C1			**C2**			**T1**			**T2**		
O···X atoms	R_1	R_2	O···X atoms	R_1	R_2	O···X atoms	R_1	R_2	O···X atoms	R_1	R_2
$O^{3'} \cdots O^{4'}$	2.43	2.86	$O^{3'} \cdots O^{4'}$	2.43	2.86	$O^{3'} \cdots O^{4'}$	2.44	2.86	$O^{3'} \cdots O^{4'}$	2.44	2.87
$O^{5'} \cdots O^{2'}$	2.46	2.88	$O^{5'} \cdots O^{2'}$	2.46	2.89						
Ouabagenin											
C1			**C2**			**T1**			**T2**		
O···X atoms	R_1	R_2	O···X atoms	R_1	R_2	O···X atoms	R_1	R_2	O···X atoms	R_1	R_2
$O^{19} \cdots O^{11}$	2.04	2.83	$O^{19} \cdots O^{11}$	2.06	2.84	$O^{1} \cdots O^{19}$	2.00	2.66	$O^{1} \cdots O^{19}$	2.00	2.66
$O^{5} \cdots O^{3}$	2.07	2.70	$O^{5} \cdots O^{3}$	2.07	2.70	$O^{19} \cdots O^{11}$	2.13	2.88	$O^{19} \cdots O^{11}$	2.15	2.89
$O^{1} \cdots O^{19}$	2.10	2.74	$O^{1} \cdots O^{19}$	2.10	2.74	$O^{11} \cdots C^{1}$	2.21	3.04	$O^{11} \cdots C^{1}$	2.20	3.04
$O^{3} \cdots O^{1}$	2.14	2.91	$O^{3} \cdots O^{1}$	2.14	2.91	$O^{5} \cdots O^{1}$	2.25	2.89	$O^{5} \cdots O^{1}$	2.27	2.91
$O^{11} \cdots C^{1}$	2.17	3.02	$O^{11} \cdots C^{1}$	2.17	3.02	$O^{5} \cdots C^{19}$	2.36	2.82	$O^{5} \cdots C^{19}$	2.35	2.81
$O^{11} \cdots C^{19}$	2.36	2.87	$O^{11} \cdots C^{19}$	2.37	2.87						

[a]Intramolecular hydrogen bonds are considered in O···H—X form, where X = O or C.

The change in ring A conformation predominantly affects the geometry and electronic parameters within this ring. In conformation family C, C^1—C^2, C^1—C^{10}, and C^4—C^5 bonds are 0.011-0.013 Å shorter, C^2—C^3 and C^5—C^{10} bonds are 0.006-0.021 Å longer, $C^1C^{10}C^9$, $C^2C^1C^{10}$, $C^3C^4C^5$, $C^4C^5C^{10}$, $C^5C^{10}C^{19}$, $C^8C^9C^{10}$, $C^9C^{10}C^{19}$, and $C^2C^1O^1$ bond angles are 0.8°-4.0° larger, $C^5C^{10}C^1$, $C^5C^{10}C^9$, $C^6C^7C^8$, $C^7C^8C^9$, $C^6C^5O^5$, and $C^5C^{10}C^{19}$ bond angles are 0.8°-2.1° smaller, the charges on C^4, O^1, O^5 atoms are less negative by 0.01-0.03 a.u., and the charges on C^1 and O^{19} atoms are larger in the absolute value by 0.01 a.u., as compared to conformation family T. The values of endocyclic torsion angles in ring B and that of $C^4C^5C^{10}C^9$ torsion angle, which describes the mutual orientation of rings A and B, are also influenced.

The rhamnosyl residue affects the structure of the steroid core mainly in conformation family C. Table **4.1** shows that the presence of the rhamnose ring causes elongation of C^3—O^3 bond by ~0.01 Å, shortening of C^5—C^{10} bond by ~0.01 Å, increase of $C^1C^{10}C^9$ bond angle by 1.6°, and decrease of $C^5C^{10}C^1$ and $C^8C^9C^{10}$ bond angles by 0.7°-1.5°. The negative charge on O^3 appreciably decreases in the absolute value. Since the presence of the rhamnosyl substituent in conformation family C alters the pattern of intramolecular hydrogen bonding involving O^3 and O^5 atoms, it is difficult to attribute the observed changes in steric and electronic parameters to the effect of a single structural factor. This mainly concerns the parameters which involve these oxygen atoms and atoms in their nearest surrounding, *e.g.*, $C^1C^2C^3$, $C^2C^3O^3$, $C^4C^3O^3$, $C^4C^5O^5$, $C^{10}C^5O^5$ bond angles, as well as the charges on C^2, C^3, C^5, C^6, C^{10}, and C^{19} atoms.

The results of our calculations demonstrate that ouabain and ouabagenin molecules can be conventionally divided into three fragments: the steroid core consisting of four fused rings, the lactone ring and the rhamnosyl residue. The steric and electronic structure of the latter two fragments remains essentially unchanged irrespective of whether they are regarded as free molecules or fragments of ouabain and ouabagenin. On the other hand, their presence as substituents in the steroid molecules affects to a certain extent the geometry and electron density distribution in the main fragment of the molecules, the steroid core. For instance, a change in orientation of the lactone ring results in distortion of ring D conformation, while introduction of the rhamnosyl residue into the ouabagenin molecule changes the pattern of intramolecular hydrogen bonding. Nevertheless, conformation of ring A is the factor that affects most strongly the steric and electronic structure of ouabain and ouabagenin.

The ring A *chair-twist* interconversion changes the distance between the oxygen atoms of the lactone substituent responsible for selective ouabain binding to Na^+,K^+-ATPase and the oxygen atoms of the rhamnosyl group providing formation of a quasi-irreversible ouabain–Na^+,K^+-ATPase complex, and this fact was suggested earlier to account for biological significance of conformational mobility of ring A [58]. It should be however noted that ouabain and ouabagenin molecules contain several oxygen atoms located on the same side of the steroid core plane at a short distance from one another. Hence, in physiological conditions these molecules can act as chelators of various positively charged ions, mainly of small radius (Ca^{2+}, Mg^{2+}, Na^+, K^+, etc.), and thus form ternary steroid–cation–Na^+,K^+-ATPase complexes upon binding to Na^+,K^+-ATPase.

Ouabain–Ca^{2+} Chelate Complexes

It is reasonable to assume that at least three oxygen atoms are involved in Ca^{2+} chelation. Analysis of ouabain steric structure makes it possible to exclude some of them from consideration. The oxygen atoms O^{14}, O^{21}, and O^{22} are too distant from each other and from other oxygen atoms and their location in space cannot be adjusted to chelate Ca^{2+} due to rather rigid structure of the ouabain molecule. The atoms $O^{2'}$, $O^{3'}$, and $O^{4'}$ are capable to simultaneously participate in Ca^{2+} chelation only if stereochemistry of the rhamnose ring is changed, moreover, they are positioned away from other oxygen atoms. Consequently, only six atoms should be considered as potential electron density donors upon formation of ouabain–Ca^{2+} complex: O^1, O^3, O^5, O^{11}, O^{19}, and $O^{5'}$.

The geometry and electronic parameters of ouabain–Ca^{2+} conformations were obtained by the following protocol. Cartesian coordinates of four stable ouabain conformations discussed in detail above were used as starting sets. All possible pairwise combinations of oxygen atoms which could be involved in Ca^{2+} chelation were considered (6 atoms, 15 combinations), and Ca^{2+} ion was placed exactly in the middle between the oxygen atoms of a given pair. 60 starting points were generated (15 positions of Ca^{2+} ion, 4 initial ouabain conformations) for further geometry optimization of ouabain–Ca^{2+} system. It is expected that in the starting point the system is quite far from the minima on the potential energy surface, but electrostatic attraction of Ca^{2+} to oxygen atoms is rather strong to keep the ion near the ouabain molecule. All stable conformations of ouabain–Ca^{2+} complexes are expected to be found according to the described procedure.

Full *ab initio* geometry optimization of all 60 starting sets of ouabain–Ca^{2+} parameters was carried out by the restricted Hartree-Fock method using 6-31G* basis set [54] and GAMESS program [55]. Six possible conformations of ouabain–Ca^{2+} complexes were obtained, and several modes of Ca^{2+} chelation were identified. In the first case, Ca^{2+} is coordinated by five oxygen atoms (O^1, O^3, O^5, O^{19}, $O^{5'}$), while ring A is in the *chair* conformation only. In the second case, the cation is bound by three oxygen atoms (O^1, O^{11}, O^{19}), and ring A can adopt either the *chair* or the *bath* conformation. It should be noted that the *twist* conformation of ring A found in the free ouabain molecule is not energetically favorable in ouabain–Ca^{2+} chelates. Regardless of the cation binding mode and ring A conformation, there are two possible orientations of the lactone ring relative to the steroid core in each of the structures. The following description of the conformations is used further: indices C and B denote the conformation of ring A (*chair* and *bath*, respectively); indices 1 and 2 are assigned to the orientation of the lactone ring ($\tau(C^{16}C^{17}C^{20}C^{21})$ values are about 150° and -45°), subscripts 3 and 5 point to the number of oxygen atoms involved in Ca^{2+} chelation.

Steric structures of $C1_5$, $C1_3$, and $B2_3$ conformations are shown in Fig. (**4.10**). Conformation $C1_5$ has the lowest energy in the gas phase, the energies of $C2_5$, $B2_3$, $B1_3$, $C2_3$, and $C1_3$ conformations being 0.8, 36.0, 37.4, 39.0, and 39.8 kcal·mol^{-1} higher, respectively. Selected values of bond lengths, bond and torsion angles, and atomic charges for all studied conformations of ouabain–Ca^{2+} chelates are listed in Table **4.4**.

Table 4.4. Selected calculated (RHF/6-31G*) bond lengths, bond and torsion angles, and atomic charges in examined conformations of ouabain–Ca^{2+} chelate complex.

Parameter	$C1_5$	$C2_5$	$C1_3$	$C2_3$	$B1_3$	$B2_3$
Bond Length, *r*, Å						
C^1—C^{10}	1.578	1.595	1.579	1.596	1.584	1.585
C^3—C^4	1.521	1.515	1.521	1.515	1.534	1.534

(Table 4.4) contd.....

Parameter	$C1_5$	$C2_5$	$C1_3$	$C2_3$	$B1_3$	$B2_3$
C^6—C^7	1.515	1.527	1.516	1.528	1.521	1.521
C^8—C^9	1.552	1.545	1.556	1.549	1.561	1.564
C^9—C^{10}	1.612	1.576	1.613	1.576	1.585	1.586
C^{10}—C^{19}	1.554	1.570	1.554	1.569	1.550	1.550
C^{12}—C^{13}	1.543	1.558	1.543	1.556	1.545	1.545
C^{13}—C^{14}	1.556	1.571	1.548	1.560	1.558	1.549
C^{13}—C^{17}	1.565	1.569	1.578	1.584	1.569	1.576
C^{14}—C^{15}	1.552	1.553	1.536	1.536	1.552	1.537
C^{15}—C^{16}	1.545	1.543	1.535	1.530	1.544	1.535
C^{16}—C^{17}	1.546	1.544	1.555	1.551	1.546	1.555
C^1—O^1	1.453	1.441	1.453	1.441	1.463	1.463
C^3—O^3	1.440	1.419	1.440	1.418	1.396	1.396
C^5—O^5	1.459	1.426	1.457	1.426	1.416	1.415
C^{11}—O^{11}	1.413	1.484	1.413	1.484	1.469	1.468
C^{19}—O^{19}	1.448	1.423	1.447	1.423	1.335	1.346
$C^{1'}$—O^3	1.409	1.395	1.410	1.396	1.396	1.397
$C^{1'}$—$O^{5'}$	1.412	1.404	1.412	1.403	1.388	1.388
$C^{5'}$—$O^{5'}$	1.453	1.435	1.453	1.435	1.429	1.429
Bond Angle, ω, deg						
$C^1C^2C^3$	113.8	115.0	113.8	115.0	109.5	109.5
$C^1C^{10}C^9$	104.2	102.6	104.4	102.5	109.6	109.7
$C^1C^{10}C^{19}$	111.0	118.3	110.9	118.2	112.0	111.9
$C^2C^1C^{10}$	116.5	116.2	116.6	116.2	118.0	117.9
$C^2C^3C^4$	108.6	107.8	108.5	107.8	110.8	110.8
$C^4C^5C^6$	109.7	107.9	109.7	107.8	108.9	108.9
$C^4C^5C^{10}$	114.1	116.0	114.2	116.0	113.4	113.4
$C^5C^{10}C^1$	112.8	112.0	112.7	112.0	107.5	107.5
$C^5C^{10}C^9$	110.3	114.7	110.3	114.6	109.1	109.0
$C^5C^{10}C^{19}$	112.0	103.8	112.1	104.0	104.7	104.8
$C^6C^5C^{10}$	113.0	111.5	113.0	111.4	112.4	112.5
$C^6C^7C^8$	110.0	111.2	110.1	111.4	112.8	112.9
$C^7C^8C^{14}$	110.0	112.2	110.9	113.0	109.2	110.0
$C^8C^9C^{10}$	116.1	116.2	116.2	116.6	113.7	114.0
$C^8C^{14}C^{15}$	114.7	114.5	116.2	116.2	114.4	116.0

(Table 4.4) contd.....

Parameter	$C1_5$	$C2_5$	$C1_3$	$C2_3$	$B1_3$	$B2_3$
$C^9C^8C^{14}$	113.9	112.2	113.5	111.7	114.4	113.9
$C^9C^{10}C^{19}$	106.1	105.7	106.0	105.8	113.6	113.7
$C^9C^{11}C^{12}$	109.9	109.6	110.5	109.6	111.9	112.5
$C^{10}C^9C^{11}$	115.4	114.8	115.1	114.8	117.5	115.2
$C^{11}C^{12}C^{13}$	114.5	111.6	114.4	111.6	113.0	112.8
$C^{12}C^{13}C^{14}$	108.7	110.7	108.1	109.8	108.7	108.0
$C^{14}C^{15}C^{16}$	107.5	107.8	104.1	103.6	107.7	104.5
$C^{16}C^{17}C^{13}$	103.6	103.5	105.7	105.9	103.5	105.5
$C^{17}C^{20}C^{21}$	122.2	122.1	126.4	125.9	122.1	126.5
$C^{17}C^{20}C^{22}$	130.3	130.3	126.0	126.5	130.3	125.9
$C^2C^1O^1$	106.9	109.5	106.9	109.5	107.2	107.2
$C^4C^5O^5$	105.9	108.7	105.9	108.7	109.2	109.2
$C^6C^5O^5$	104.7	107.9	104.7	107.8	109.1	108.9
$C^8C^{14}O^{14}$	102.0	102.2	107.5	108.1	102.1	107.9
$C^9C^{11}O^{11}$	111.6	114.8	111.4	114.6	114.5	114.2
$C^{10}C^1O^1$	112.7	115.8	112.6	115.8	112.2	112.2
$C^{10}C^5O^5$	108.7	104.5	108.7	104.8	103.7	103.8
$C^{10}C^{19}O^{19}$	116.6	118.1	116.7	118.1	115.2	115.2
$C^{12}C^{11}O^{11}$	109.7	110.6	109.8	110.8	107.5	107.5
$C^{13}C^{14}O^{14}$	111.0	110.4	106.3	105.3	110.3	105.4
$C^{15}C^{14}O^{14}$	110.8	111.1	109.2	109.0	111.3	109.3
$C^3O^3C^{1'}$	121.5	118.3	121.4	118.3	118.2	118.1
$C^{1'}C^{2'}C^{3'}$	112.4	111.3	112.3	111.3	110.6	110.7
$C^{2'}C^{1'}O^{5'}$	112.8	112.1	112.8	112.2	111.2	111.3
$C^{1'}O^{5'}C^{5'}$	118.4	117.4	118.5	117.4	116.7	116.8
$O^3C^{1'}C^2$	115.6	111.0	115.5	111.0	108.3	108.3
$O^3C^{1'}O^{5'}$	103.3	108.7	103.3	108.7	112.4	112.5
Torsion Angle, τ, deg						
$C^1C^2C^3C^4$	-57.3	-58.1	-57.4	-58.1	75.7	57.5
$C^2C^3C^4C^5$	57.6	58.3	57.9	58.3	-11.5	-10.8
$C^3C^4C^5C^{10}$	-47.0	-47.7	-47.3	-47.9	-43.4	-43.8
$C^4C^5C^{10}C^1$	33.7	32.3	33.6	32.4	48.0	47.6
$C^5C^{10}C^1C^2$	-35.4	-32.2	-35.1	-32.2	-1.3	-0.5
$C^{10}C^1C^2C^3$	48.7	47.2	48.5	47.3	-52.2	-52.7

(Table 4.4) contd.....

Parameter	$C1_5$	$C2_5$	$C1_3$	$C2_3$	$B1_3$	$B2_3$
$C^5C^6C^7C^8$	60.7	61.1	60.9	61.4	52.9	53.5
$C^6C^7C^8C^9$	-54.3	-54.7	-54.5	-54.4	-49.0	-49.4
$C^7C^8C^9C^{10}$	46.1	43.7	46.3	43.3	49.0	49.0
$C^8C^9C^{10}C^5$	-39.7	-36.7	-39.9	-36.7	-50.0	-49.8
$C^9C^{10}C^5C^6$	43.8	39.9	43.7	40.0	53.3	52.9
$C^{10}C^5C^6C^7$	-56.7	-52.7	-56.5	-53.0	-56.1	-56.1
$C^8C^9C^{11}C^{12}$	62.2	68.6	60.6	67.8	61.9	60.3
$C^9C^{11}C^{12}C^{13}$	-63.0	-64.6	-62.4	-65.2	-64.3	-63.5
$C^{11}C^{12}C^{13}C^{14}$	51.5	49.1	53.1	51.1	52.1	53.6
$C^{12}C^{13}C^{14}C^8$	-45.2	-42.5	-48.1	-45.1	-45.4	-48.6
$C^{13}C^{14}C^8C^9$	52.5	51.2	54.5	52.8	50.6	53.0
$C^{14}C^8C^9C^{11}$	-58.4	-61.5	-57.9	-61.2	-55.4	-55.1
$C^{13}C^{14}C^{15}C^{16}$	16.6	12.6	40.2	41.8	17.0	39.2
$C^{14}C^{15}C^{16}C^{17}$	6.6	10.9	-29.0	-34.5	6.0	-27.4
$C^{15}C^{16}C^{17}C^{13}$	-26.9	-29.9	6.8	13.8	-26.3	5.1
$C^{16}C^{17}C^{13}C^{14}$	37.1	37.4	17.7	11.7	36.8	18.9
$C^{17}C^{13}C^{14}C^{15}$	-33.1	-30.8	-35.6	-32.7	-33.2	-35.6
$C^4C^5C^{10}C^9$	-82.4	-84.2	-82.6	-83.9	-70.8	-71.1
$C^7C^8C^9C^{11}$	174.8	171.2	174.9	171.1	178.3	178.3
$C^{12}C^{13}C^{14}C^{15}$	80.8	83.4	78.9	82.1	80.5	78.4
$C^{16}C^{17}C^{20}C^{21}$	151.9	151.7	-47.4	-41.2	152.3	-46.5
$C^{16}C^{17}C^{20}C^{22}$	-22.5	-22.4	130.5	137.2	-21.6	131.5
Atomic Charge, q, a.u.						
C^1	0.11	0.16	0.11	0.16	0.18	0.18
C^2	-0.37	-0.42	-0.37	-0.42	-0.42	-0.42
C^3	0.13	0.16	0.13	0.16	0.21	0.21
C^4	-0.33	-0.40	-0.34	-0.40	-0.36	-0.36
C^5	0.31	0.34	0.31	0.34	0.20	0.20
C^6	-0.37	-0.34	-0.37	-0.34	-0.36	-0.35
C^9	-0.23	-0.24	-0.23	-0.25	-0.21	-0.21
C^{12}	-0.37	-0.40	-0.37	-0.40	-0.38	-0.39
C^{13}	-0.08	-0.08	-0.05	-0.05	-0.08	-0.05
C^{14}	0.38	0.38	0.35	0.34	0.38	0.35
C^{19}	0.01	-0.08	0.01	-0.08	0.02	0.02

(Table 4.4) contd.....

Parameter	C1$_5$	C2$_5$	C1$_3$	C2$_3$	B1$_3$	B2$_3$
C^{22}	-0.39	-0.39	-0.34	-0.34	-0.39	-0.34
O^1	-0.88	-0.96	-0.88	-0.96	-0.92	-0.92
O^3	-0.77	-0.72	-0.77	-0.72	-0.68	-0.68
O^5	-0.89	-0.84	-0.88	-0.84	-0.79	-0.79
O^{11}	-0.81	-0.93	-0.80	-0.93	-0.93	-0.93
O^{19}	-0.88	-0.85	-0.86	-0.85	-0.91	-0.91
C$^{1'}$	0.47	0.43	0.47	0.43	0.43	0.43
C$^{2'}$	0.08	0.11	0.08	0.11	0.11	0.11
C$^{5'}$	0.08	0.13	0.08	0.13	0.14	0.15
C$^{6'}$	-0.50	-0.47	-0.50	-0.47	-0.48	-0.48
O$^{5'}$	-0.77	-0.73	-0.77	-0.72	-0.69	-0.69
Ca	1.56	1.56	1.67	1.67	1.69	1.69

(a)

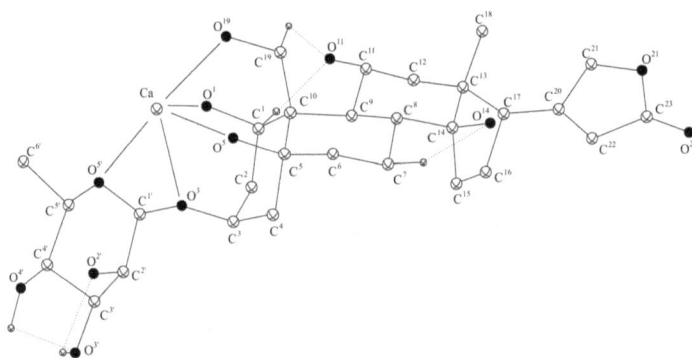

(b)

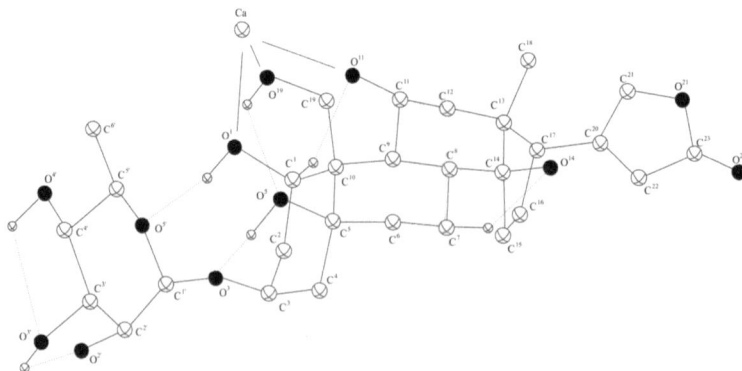

Fig. 4.10 contd.....

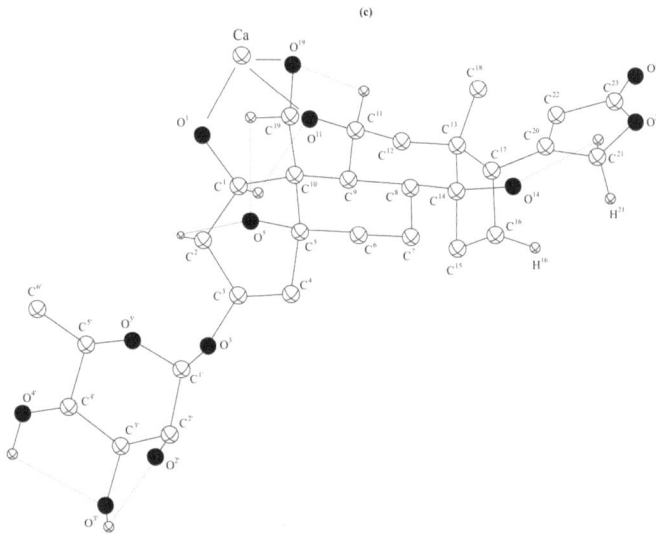

Fig. (4.10). Steric structures of selected ouabain–Ca^{2+} conformations, accepted atom numbering, and intramolecular hydrogen bonding patterns:
a – Conformation $C1_5$;
b – Conformation $C1_3$;
c – Conformation $B2_3$.
Only some of hydrogen atoms are shown. Hydrogen bonds are presented as dotted lines; coordination Ca–O bonds, as arrows.

Comparing the data presented in Tables **4.1**, **4.2** and **4.4**, the effect of Ca^{2+} chelation on molecular structure of ouabain can be analyzed. The lengths of C—O bonds which involve the oxygen atoms participating in chelation increase by 0.02-0.08 Å, while C^{14}—O^{14}, $C^{2'}$—$O^{2'}$, $C^{3'}$—$O^{3'}$, and $C^{4'}$—$O^{4'}$ bonds shorten by 0.01-0.02 Å. Endocyclic bonds within ring A extend by 0.01-0.03 Å. The structure of ring D in conformation family 1 is changed: C^{13}—C^{17} bond becomes shorter, while C^{13}—C^{14}, C^{14}—C^{15}, and C^{15}—C^{16} bonds elongate by 0.01-0.02 Å. These variations in bond lengths are interconnected with the changes of bond angles involving the oxygen atoms bound to Ca^{2+}, as well as of endocyclic bond angles within ring A. The values of ring B endocyclic bond angles are also affected, though noticeably less. Bond angles involving the atoms of ring C remain fairly unchanged, except for $C^{11}C^{12}C^{13}$ bond angle, the value of which is reduced by 3° in conformation family 3. Bond angles within ring D and those involving O^{14} oxygen atom are substantially affected in conformation family 5.

Planarity of the lactone rings is slightly distorted upon Ca^{2+} chelation: the absolute values of torsion angles within these rings $|\tau| \leq 3.0°$, whereas in free ouabain $|\tau| \leq 0.5°$. Ring A becomes noticeably more planar: the simple arithmetic average of the absolute values of torsion angles within this ring (τ_{av}) is 46.0°-46.6°, which is by 5.5°-6.0° less than in free ouabain. Binding of Ca^{2+} results in the decrease of τ_{av} for ring B by ~3.5° in $C1_3$ and $C2_3$ conformations and by ~1.5° in $C1_5$ and $C2_5$ conformations, the corresponding τ_{av} values being equal to 48.1° and 50.2°. The values of τ_{av} for ring C in all examined conformations and for ring B in conformations $B1_3$ and $B2_3$ remain practically unchanged. The rhamnose rings also become more planar upon Ca^{2+} chelation: τ_{av} values for these rings are by 1.5°-4.0° less as compared to free ouabain. Conformation of ring D is significantly affected. In contrast to the 14β-*envelope* conformation observed in free ouabain, this ring adopts the *half-chair* conformation in ouabain–Ca^{2+} complexes. Moreover, the shapes of the *half-chair* differ between conformation families 1 and 2. In conformation family 1, atoms C^{13} and C^{17} are located on opposite sides of the plane formed by C^{14}, C^{15}, and C^{16} carbon atoms, while in conformation family 2 atoms C^{14} and C^{15} are located on opposite sides of the plane formed by C^{13}, C^{16}, and C^{17}. The *half-chair* is distorted toward 13α-*envelope* in conformations $C1_5$ and $B1_3$; toward 14β-*envelope*, in conformations $C2_5$ and $B2_3$; and it is fairly undistorted in conformations $C1_3$ and $C2_3$ (Table **4.4**). Ca^{2+} chelation considerably influences $\tau(C^4C^5C^{10}C^9)$, determining mutual orientation of rings A and B, and the values of $\tau(C^{16}C^{17}C^{20}C^{21})$ and $\tau(C^{16}C^{17}C^{20}C^{22})$ defining location of the lactone ring relative to the steroid core. Mutual orientation of rings B and C, as well as of rings C and D, remains unchanged.

The most significant changes in Mulliken atomic charges upon chelation are observed for the oxygen atoms bound to Ca^{2+}: the charges become more negative by 0.06-0.15 a.u. The absolute values of charges on O^{23}, $O^{2'}$, and $O^{3'}$ decrease by 0.02-0.04 a.u. In conformation family 1, the charges on ring D carbon atoms C^{13}, C^{15}, and C^{16} shift towards more negative values by 0.02-0.04 a.u., while the positive charge on C^{14} increases by 0.04-0.05 a.u. The charge on C^{22} atom of the lactone ring, which is linked to ring D, becomes more negative by 0.05 a.u. A significant charge redistribution is detected for ring A atoms. In conformation family 5, C^1, C^2, C^3, and C^{10} atoms are charged more negatively by 0.02-0.05 a.u., whereas the charge on C^4 decreases by 0.04 a.u. in the absolute value. In conformation family 3, the charges on C^2 and C^4 atoms grow in the absolute value by 0.05-0.07 a.u. and 0.02-0.03 a.u., correspondingly; the charges on C^3 and C^{10} atoms become more positive by 0.02-0.03 a.u. The charges on atoms of rings B and C remain almost unaffected, except for the charges on C^6 and C^{12} atoms which are shifted by 0.03-0.07 a.u. towards more negative values.

The following explanation of the changes in steric and electronic structure of ouabain upon Ca^{2+} chelation may be proposed. Involvement of O^1, O^3, O^5, O^{11}, O^{19}, and $O^{5'}$ atoms in coordination of Ca^{2+} accounts for weakening of the corresponding C—O bonds, which also affects the values of respective bond angles and charges on these oxygen atoms, as well as on the carbon atoms in their vicinity. The most pronounced changes in geometry and electronic parameters are observed in ring A, since in all studied ouabain–Ca^{2+} conformations this ring contains the hydroxyl groups which donate electron density upon complex formation. Structural changes in the rhamnosyl ring and ring C occur mainly in conformation families 5 and 3, correspondingly, because $O^{5'}$ atom of the rhamnose is involved in Ca^{2+} chelation only in conformation family 5, while O^{11} atom bonded to C^{11} atom of ring C interacts with the cation exclusively in conformation family 3.

The changes in ring D conformation can be most easily illustrated by analyzing intramolecular hydrogen bonding patterns presented in Table **4.5**. The total number of hydrogen bonds in ouabain–Ca^{2+} complexes is less than that in conformations of free ouabain. Approximately a half of them are of $O \cdots H - C$ type, as oxygen atoms are mostly involved in Ca^{2+} chelation. Some of $O \cdots H - C$ hydrogen bonds were detected to be rather strong, *e.g.*, $O^{11} \cdots H - C^1$ bonds in conformations $C1_5$ and $C2_5$. All free ouabain and ouabain–Ca^{2+} conformations of family 2 are stabilized by weak $O^{14} \cdots H - C^{21}$ hydrogen bonds and by attraction of H^{16} and H^{21} hydrogen atoms (Fig. **4.10.c**). In ouabain–Ca^{2+} chelates belonging to conformation family 1, $O^{14} \cdots H - C^7$ hydrogen bonds are formed instead of $O^{14} \cdots H - C^{22}$ bonds found in free ouabain conformations of this family, which accounts for significant changes in ring D conformation upon Ca^{2+} chelation observed in this conformation family.

Table **4.5**. Calculated parameters of intramolecular hydrogen bonds[a] present in examined conformations of ouabain–Ca^{2+} chelate complex: R_1 (Å), $O \cdots H$ distance; R_2 (Å), $O \cdots X$ distance.

$C1_5$			$C2_5$			$C1_3$		
$O \cdots X$ atoms	R_1	R_2	$O \cdots X$ atoms	R_1	R_2	$O \cdots X$ atoms	R_1	R_2
$O^{11} \cdots C^1$	1.98	2.84	$O^{11} \cdots C^1$	1.97	2.84	$O^{5'} \cdots O^1$	1.84	2.81
$O^{2'} \cdots O^{3'}$	2.34	2.80	$O^{2'} \cdots O^{3'}$	2.35	2.80	$O^{5'} \cdots O^{19}$	2.07	2.71
$O^{11} \cdots C^{19}$	2.35	3.10	$O^{11} \cdots C^{19}$	2.37	3.11	$O^3 \cdots O^5$	2.13	2.76
$O^{14} \cdots C^7$	2.36	2.75	$O^{3'} \cdots O^{4'}$	2.48	2.86	$O^{11} \cdots C^1$	2.22	2.92
$O^{3'} \cdots O^{4'}$	2.48	2.86	$O^{14} \cdots C^{21}$	2.70	3.29	$O^{2'} \cdots O^{3'}$	2.33	2.80
						$O^{14} \cdots C^7$	2.46	2.82
						$O^{3'} \cdots O^{4'}$	2.48	2.86
$C2_3$			$B1_3$			$B2_3$		
$O \cdots X$ atoms	R_1	R_2	$O \cdots X$ atoms	R_1	R_2	$O \cdots X$ atoms	R_1	R_2
$O^{5'} \cdots O^1$	1.85	2.81	$O^{5'} \cdots C^{19}$	2.24	2.60	$O^{5'} \cdots C^{19}$	2.24	2.60
$O^{5'} \cdots O^{19}$	2.10	2.72	$O^{5'} \cdots C^2$	2.33	2.84	$O^{5'} \cdots C^2$	2.33	2.84
$O^3 \cdots O^5$	2.11	2.77	$O^{2'} \cdots O^{3'}$	2.33	2.80	$O^{2'} \cdots O^{3'}$	2.33	2.80

(Table 4.5) contd.....

C1$_5$			C2$_5$			C1$_3$		
O···X atoms	R_1	R_2	O···X atoms	R_1	R_2	O···X atoms	R_1	R_2
O^{11}···C^1	2.21	2.92	O^{11}···C^1	2.34	2.97	O^{11}···C^1	2.34	2.96
O$^{2'}$···O$^{3'}$	2.33	2.80	O^{19}···C^{11}	2.37	2.93	O^{19}···C^{11}	2.35	2.93
O$^{3'}$···O$^{4'}$	2.48	2.86	O^{14}···C^7	2.39	2.76	O$^{3'}$···O$^{4'}$	2.48	2.87
O^{14}···C^{21}	2.88	3.51	O$^{3'}$···O$^{4'}$	2.48	2.87	O^{14}···C^{21}	2.64	3.25

[a]Intramolecular hydrogen bonds are considered in O···H—X form, where X = O or C.

The values of distances r(Ca—O) in optimized structures of ouabain–Ca^{2+} chelates do not differ between conformation families 1 and 2. In C1$_5$ conformation: r(Ca—O^1) = 2.34 Å, r(Ca—O^3) = 2.49 Å, r(Ca—O^5) = 2.34 Å, r(Ca—O^{19}) = 2.40 Å, r(Ca—O$^{5'}$) = 2.47 Å; in C1$_3$ conformation: r(Ca—O^1) = 2.28 Å, r(Ca—O^{11}) = 2.48 Å, r(Ca—O^{19}) = 2.33 Å; in B1$_3$ conformation: r(Ca—O^1) = 2.30 Å, r(Ca—O^{11}) = 2.33 Å, r(Ca—O^{19}) = 2.32 Å. Distances r(Ca—O^1) and r(Ca—O^{19}) in conformation family 5 are larger than those in conformation family 3 by 0.04-0.08 Å, because of rigidity of the steroid core and electrostatic repulsion between the chelating oxygen atoms. Increase of r(Ca—O^{11}) in conformation C1$_3$ by 0.15 Å as compared to that in conformation B1$_3$ is accounted for by the presence in the former conformation of strong O^5···H—O^{19} hydrogen bond fixating the steric location of O^{19} atom. As a result, it is more energetically favorable for the cation to move away from O^{11} atom, while being rather tightly bound to O^{19} atom.

The values of steric and electronic parameters which remain fairly unchanged in six examined ouabain–Ca^{2+} chelates are presented in Table **4.6** for C1$_5$ conformation. They are mostly related to the structure of the lactone and rhamnosyl rings, which supports the earlier observation that these rings may be considered as independent fragments, weakly affected by the steroid core of ouabain.

The results of our calculations demonstrate that the ouabain molecule is theoretically capable of chelating Ca^{2+}. The process of chelation noticeably affects the steric and electronic structure of ouabain. The most pronounced changes are observed in rings A and D of the steroid core, and they are accounted for by involvement of oxygen atoms in formation of Ca—O coordination bonds and by rearrangement of intramolecular hydrogen bonding patterns upon chelation.

In accordance with currently accepted models of binding of cardiac glycosides to Na$^+$,K$^+$-ATPase, the main contribution to the energy of formation of ouabain–Na$^+$,K$^+$-ATPase complex is provided by numerous weak van der Waals interactions and, to a lesser extent, by formation of intermolecular hydrogen bonds [59, 60]. It should be emphasized that these models describe the molecular mechanism of modulation of the Na^+,K^+-ATPase pumping function by cardiotonic steroids. Such a mechanism is consistent with relatively high acting concentrations of the glycosides (10^{-4}–10^{-5} M), which suggests an absence of strong specific ion-ionic interactions between the glycoside molecule and Na^+,K^+-ATPase. The driving force of the process in this case is the diffusion effect which facilitates penetration of attacking molecules into the binding site responsible for inhibition of the Na^+,K^+-ATPase pumping function when their concentration exceeds the threshold level.

Table 4.6. Selected calculated (RHF/6-31G*) bond lengths, bond angles, and atomic charges in conformation $C1_s$ of ouabain–Ca^{2+} chelate complex.

Bond Length, r, Å									
C^1—C^2	1.525	C^8—C^{14}	1.553	C^{20}—C^{21}	1.514	C^{14}—O^{14}	1.406	$C^{2'}$—$O^{2'}$	1.399
C^2—C^3	1.518	C^9—C^{11}	1.552	C^{20}—C^{22}	1.324	$C^{1'}$—$C^{2'}$	1.527	$C^{3'}$—$O^{3'}$	1.388
C^4—C^5	1.541	C^{11}—C^{12}	1.527	C^{22}—C^{23}	1.485	$C^{2'}$—$C^{3'}$	1.525	$C^{4'}$—$O^{4'}$	1.387
C^5—C^6	1.529	C^{13}—C^{18}	1.535	C^{21}—O^{21}	1.403	$C^{3'}$—$C^{4'}$	1.525		
C^5—C^{10}	1.590	C^{12}—C^{13}	1.542	C^{23}—O^{21}	1.343	$C^{4'}$—$C^{5'}$	1.523		
C^7—C^8	1.526	C^{17}—C^{20}	1.514	C^{23}—O^{23}	1.177	$C^{5'}$—$C^{6'}$	1.516		
Bond Angle, ω, deg									
$C^3C^4C^5$	115.8	$C^{13}C^{14}C^{15}$	104.6	$C^{20}C^{22}C^{23}$	109.2	$C^{21}O^{21}C^{23}$	111.1	$C^{3'}C^{2'}O^{2'}$	111.7
$C^5C^6C^7$	113.7	$C^{13}C^{17}C^{20}$	115.9	$C^{21}C^{20}C^{22}$	107.3	$O^{21}C^{23}O^{23}$	124.1	$C^{3'}C^{4'}O^{4'}$	111.3
$C^7C^8C^9$	113.2	$C^{14}C^{13}C^{17}$	103.9	$C^2C^3O^3$	112.7	$C^{2'}C^{3'}C^{4'}$	110.8	$C^{4'}C^{3'}O^{3'}$	110.8
$C^8C^9C^{11}$	105.0	$C^{14}C^{13}C^{18}$	113.2	$C^4C^3O^3$	108.1	$C^{3'}C^{4'}C^{5'}$	109.6	$C^{4'}C^{5'}O^{5'}$	110.6
$C^{12}C^{13}C^{17}$	107.1	$C^{15}C^{16}C^{17}$	106.7	$C^{20}C^{21}O^{21}$	105.1	$C^{4'}C^{5'}C^{6'}$	113.9	$C^{5'}C^{4'}O^{4'}$	105.7
$C^{12}C^{13}C^{18}$	109.5	$C^{16}C^{17}C^{20}$	112.7	$C^{22}C^{23}O^{21}$	107.2	$C^{1'}C^{2'}O^{2'}$	109.2	$C^{6'}C^{5'}O^{5'}$	106.8
$C^{13}C^{14}C^8$	114.0	$C^{17}C^{13}C^{18}$	114.1	$C^{22}C^{23}O^{23}$	128.7	$C^{2'}C^{3'}O^{3'}$	111.8		
Atomic Charge, q, a.u.									
C^7	-0.34	C^{16}	-0.34	C^{21}	0.01	$C^{3'}$	0.15	$O^{4'}$	-0.75
C^8	-0.19	C^{17}	0.21	C^{23}	0.83	$C^{4'}$	0.15		
C^{11}	0.20	C^{18}	-0.48	O^{21}	-0.61	$O^{2'}$	-0.75		
C^{15}	-0.37	C^{20}	0.01	O^{23}	-0.53	$O^{3'}$	-0.75		

However, experimentally observed ability of ouabain to modulate the transducing function of Na^+,K^+-ATPase at nanomolar concentrations indicates the presence of specific intermolecular interactions upon binding of ouabain to the site controlling the functioning of Na^+,K^+-ATPase as the signal transducer. We suggest that ouabain binding to this site proceeds *via* formation of ternary ouabain–Ca^{2+}–Na^+,K^+-ATPase complex, where Ca^{2+} forms intermolecular ion-ionic bonds between attacking ouabain–Ca^{2+} molecule and Na^+,K^+-ATPase, while the pumping Na^+,K^+-ATPase function is inhibited by free ouabain.

Ouabagenin, despite being structurally close to ouabain and potentially capable of chelating Ca^{2+}, was shown to affect the functioning of $Na_V1.8$ channels by another mechanism, which does not necessarily involve Na^+,K^+-ATPase.

Ouabagenin–Ca^{2+} Chelate Complexes

From the viewpoint of chemical structure, ouabagenin is the aglycone of ouabain, *i.e.*, their molecules differ only by presence or absence of the rhamnosyl substituent in position 3β of the steroid core. Ouabagenin contains eight oxygen atoms; under physiological conditions, it should be capable of chelating various cations, including Ca^{2+} ions. Similar to ouabain (see above), it is reasonable to assume participation of at least three oxygen atoms in Ca^{2+} chelation, and only five out of eight oxygen atoms (O^1, O^3, O^5, O^{11}, and O^{19}) may be considered as potential electron density donors.

The calculation procedure applied was the same as used for investigation of ouabain–Ca^{2+} chelates. The number of starting sets of ouabagenin–Ca^{2+} geometry parameters was equal to 40 (10 positions of Ca^{2+} ions and 4 possible ouabagenin conformations); they were fully optimized *ab initio* by the restricted Hartree–Fock method with the 6-31G* basis set [54] using GAMESS program [55]. As a result, eight conformations of ouabagenin–Ca^{2+} complexes were obtained, which differed by the mode of cation binding. In the first case, Ca^{2+} ion is coordinated to four oxygen atoms (O^1, O^3, O^5, O^{19}), and the only stable conformation of ring A is *chair*. In the second case, the cation is bound to three oxygen atoms (O^1, O^3, O^5), and ring A also adopts exclusively the *chair* conformation. In the third structure, Ca^{2+} is chelated by another three oxygen atoms (O^1, O^{11}, O^{19}), while ring A can have either *chair* or *twist* conformations. Regardless of coordination mode and ring A conformation, there are two possible orientations of the lactone ring with respect to the steroid core in each of the above structures. Further we follow classification used previously, according to which the subscripts 3 and 4 indicate the number of oxygen atoms binding Ca^{2+}. The subscript 3A refers to the conformation in which the cation is coordinated by O^1, O^3, and O^5 atoms in order to distinguish it from those where Ca^{2+} ion is coordinated by other three oxygen atoms (O^1, O^{11}, O^{19}). The indices C and T denote the conformation of ring A (*chair* and *twist*, respectively); indices 1 and 2 are assigned to the orientation of the lactone ring ($\tau(C^{16}C^{17}C^{20}C^{21})$ values are about 150° and -45°).

The structures of $C2_4$, $C1_{3A}$, $C2_3$, and $T2_3$ conformations are shown in Fig. (**4.11**). Among these, $C2_4$ is characterized by the lowest energy in the gas phase. The energies of $C1_4$, $C1_{3A}$, $C2_{3A}$, $T2_3$, $T1_3$, $C2_3$, and $C1_3$ conformations are higher by 0.5, 8.5, 9.7, 24.3, 26.0, 27.2, and 28.2 kcal·mol^{-1}, respectively. Selected bond lengths, bond angles, torsion angles, and atomic charges in all examined conformations of ouabagenin–Ca^{2+} complexes are presented in Table **4.7**.

Table 4.7. Selected calculated (RHF/6-31G*) bond lengths, bond and torsion angles, and atomic charges in examined conformations of ouabagenin–Ca^{2+} chelate complex.

Parameter	$C1_4$	$C2_4$	$C1_{3A}$	$C2_{3A}$	$C1_3$	$C2_3$	$T1_3$	$T2_3$
Bond Length, r, Å								
C^1—C^{10}	1.583	1.581	1.563	1.563	1.587	1.588	1.586	1.586
C^2—C^3	1.514	1.513	1.519	1.519	1.512	1.512	1.539	1.539
C^5—C^{10}	1.590	1.591	1.572	1.572	1.584	1.583	1.591	1.590
C^9—C^{10}	1.611	1.610	1.591	1.592	1.578	1.578	1.596	1.596
C^{13}—C^{14}	1.552	1.548	1.559	1.549	1.575	1.560	1.555	1.548
C^{13}—C^{17}	1.535	1.536	1.553	1.536	1.535	1.536	1.550	1.536
C^{14}—C^{15}	1.593	1.578	1.569	1.578	1.591	1.585	1.570	1.578
C^{16}—C^{17}	1.531	1.535	1.544	1.535	1.529	1.530	1.545	1.534
C^1—O^1	1.454	1.456	1.460	1.460	1.440	1.440	1.470	1.470
C^3—O^3	1.459	1.458	1.467	1.467	1.415	1.415	1.403	1.403
C^5—O^5	1.461	1.461	1.487	1.486	1.428	1.427	1.410	1.410
C^{11}—O^{11}	1.414	1.413	1.416	1.415	1.485	1.486	1.469	1.469
C^{19}—O^{19}	1.449	1.452	1.407	1.406	1.425	1.424	1.471	1.470
Bond Angle, ω, deg								
$C^1C^2C^3$	114.0	114.4	115.8	115.7	111.7	111.8	115.2	115.1
$C^1C^{10}C^9$	104.0	104.4	109.7	109.9	102.3	102.3	110.0	110.0
$C^1C^{10}C^{19}$	111.1	110.3	108.4	108.2	118.0	118.0	112.0	112.0
$C^2C^1C^{10}$	116.8	116.9	116.2	116.2	115.5	115.6	118.2	118.1
$C^2C^3C^4$	109.2	109.4	111.5	111.5	108.1	108.0	110.9	110.8
$C^3C^4C^5$	115.6	115.5	116.8	116.8	116.1	116.1	112.4	112.3
$C^4C^5C^6$	110.0	109.4	109.9	109.9	108.4	108.4	110.6	110.6
$C^4C^5C^{10}$	114.3	114.5	113.4	113.3	116.5	116.5	109.8	109.7
$C^5C^{10}C^1$	113.1	112.8	109.1	109.1	111.5	111.5	106.6	106.5
$C^5C^{10}C^9$	110.2	110.7	109.5	109.5	115.1	115.0	109.1	109.0
$C^5C^{10}C^{19}$	111.8	111.9	110.4	110.5	104.6	104.7	104.2	104.3

(Table 4.7) contd.....

Parameter	$C1_4$	$C2_4$	$C1_{3A}$	$C2_{3A}$	$C1_3$	$C2_3$	$T1_3$	$T2_3$
$C^7C^8C^9$	113.0	112.9	114.6	114.0	111.6	111.4	112.5	112.0
$C^8C^9C^{10}$	116.5	116.1	114.1	114.1	116.1	116.6	114.1	114.4
$C^8C^{14}C^{15}$	116.4	116.3	114.7	116.3	116.1	116.3	114.4	115.9
$C^9C^8C^{14}$	113.6	113.4	113.6	113.3	112.7	111.9	115.0	114.6
$C^9C^{10}C^{19}$	106.2	106.2	109.7	109.6	105.6	105.7	114.5	114.6
$C^9C^{11}C^{12}$	110.1	110.4	109.9	110.6	108.8	109.4	112.6	112.9
$C^{10}C^9C^{11}$	115.0	114.9	115.0	114.7	115.0	114.9	117.7	117.5
$C^{11}C^{12}C^{13}$	114.5	114.5	114.3	114.2	111.9	111.7	113.1	113.1
$C^{14}C^{15}C^{16}$	103.4	104.1	107.6	104.2	103.5	103.5	107.6	104.2
$C^{16}C^{17}C^{13}$	105.8	105.7	103.5	105.7	105.9	105.9	103.6	105.5
$C^{17}C^{20}C^{21}$	121.4	126.4	122.2	126.4	121.3	125.9	122.1	126.4
$C^{17}C^{20}C^{22}$	130.8	126.0	130.2	126.0	131.0	126.4	130.3	126.0
$C^2C^1O^1$	107.1	107.5	108.6	108.6	108.6	108.6	104.6	104.7
$C^2C^3O^3$	110.5	109.9	110.2	110.3	106.0	106.1	109.9	109.9
$C^4C^3O^3$	111.3	111.7	111.2	111.1	113.9	113.8	107.2	107.1
$C^4C^5O^5$	105.1	103.8	105.7	105.7	107.5	107.8	110.0	110.1
$C^6C^5O^5$	104.9	105.1	105.7	105.7	109.6	109.4	108.3	108.0
$C^8C^{14}O^{14}$	106.8	107.4	102.2	107.6	102.3	107.9	102.1	107.8
$C^9C^{11}O^{11}$	111.5	111.3	111.1	110.9	114.7	114.6	113.5	113.4
$C^{10}C^1O^1$	112.5	111.8	109.9	109.9	115.3	115.3	112.9	112.9
$C^{10}C^5O^5$	108.7	110.1	109.5	109.5	103.1	103.1	104.9	105.2
$C^{10}C^{19}O^{19}$	116.4	116.2	113.1	113.0	117.6	117.6	115.7	115.7
$C^{12}C^{11}O^{11}$	109.9	109.8	110.1	110.3	110.6	110.6	107.3	107.4
$C^{13}C^{14}O^{14}$	106.6	106.4	111.0	106.3	110.3	105.5	110.3	105.3
Torsion Angle, τ, deg								
$C^1C^2C^3C^4$	−56.6	−55.6	−44.5	−44.6	−62.2	−62.3	23.9	24.1
$C^2C^3C^4C^5$	57.7	57.2	46.7	46.7	55.5	55.6	33.8	33.9
$C^3C^4C^5C^{10}$	−46.9	−47.2	−50.2	−50.3	−40.9	−40.1	−72.3	−72.5
$C^4C^5C^{10}C^1$	32.8	33.5	47.6	47.7	301	29.8	44.4	44.3
$C^5C^{10}C^1C^2$	−33.5	−33.6	−47.5	−47.7	−38.0	−37.6	12.3	12.6
$C^{10}C^1C^2C^3$	46.7	46.0	47.8	48.0	56.1	55.9	−49.6	−50.0
$C^5C^6C^7C^8$	61.1	61.0	56.9	57.1	62.2	62.2	57.5	57.7
$C^6C^7C^8C^9$	−54.2	−54.8	−50.0	−50.7	−56.1	−55.5	−54.1	−54.0
$C^7C^8C^9C^{10}$	45.6	46.0	46.0	46.9	44.3	43.7	50.8	50.6

(Table 4.7) contd.....

Parameter	$C1_4$	$C2_4$	$C1_{3A}$	$C2_{3A}$	$C1_3$	$C2_3$	$T1_3$	$T2_3$
$C^8C^9C^{10}C^5$	−39.3	−39.2	−45.9	−46.5	−36.5	−36.1	−47.6	−47.3
$C^9C^{10}C^5C^6$	43.8	43.1	52.7	52.6	39.1	38.9	49.7	49.7
$C^{10}C^5C^6C^7$	−56.9	−56.0	−59.9	−59.5	−52.4	−52.5	−56.5	−56.6
$C^8C^9C^{11}C^{12}$	61.7	60.5	63.5	61.2	69.0	67.5	59.3	58.5
$C^9C^{11}C^{12}C^{13}$	−62.7	−61.9	−62.9	−62.4	−65.9	−65.2	−63.8	−63.3
$C^{11}C^{12}C^{13}C^{14}$	52.1	52.9	50.4	52.7	49.2	51.4	53.3	54.0
$C^{12}C^{13}C^{14}C^8$	−46.5	−48.2	−44.4	−48.1	−40.6	−45.0	−46.5	−48.1
$C^{13}C^{14}C^8C^9$	53.7	54.8	52.8	54.8	48.8	52.4	49.7	51.0
$C^{14}C^8C^9C^{11}$	−58.6	−58.1	−59.8	−58.6	−60.6	−60.9	−52.6	−52.6
$C^{13}C^{14}C^{15}C^{16}$	42.2	40.2	15.2	40.0	41.6	42.0	19.1	40.1
$C^{14}C^{15}C^{16}C^{17}$	−34.0	−29.0	8.1	−28.6	−38.9	−35.1	3.5	−29.6
$C^{15}C^{16}C^{17}C^{13}$	12.8	6.8	−28.1	6.4	20.8	14.5	−24.4	7.7
$C^{16}C^{17}C^{13}C^{14}$	12.9	17.7	37.4	17.9	4.6	11.1	36.2	16.9
$C^{17}C^{13}C^{14}C^{15}$	−33.7	−35.6	−32.4	−35.6	−28.1	−32.5	−34.1	−35.0
$C^4C^5C^{10}C^9$	−83.1	−83.1	−72.5	−72.6	−85.9	−86.0	−74.3	−74.3
$C^{10}C^9C^8C^{14}$	172.9	173.4	173.7	175.1	171.6	171.2	176.8	176.8
$C^{12}C^{13}C^{14}C^{15}$	80.9	78.8	81.5	78.9	87.0	82.2	79.4	79.0
$C^{16}C^{17}C^{20}C^{21}$	152.7	−47.6	152.0	−47.1	146.6	−41.3	152.8	−45.7
Atomic Charge, q, a.u.								
C^1	0.10	0.09	0.13	0.13	0.09	0.09	0.20	0.20
C^2	−0.35	−0.34	−0.38	−0.38	−0.37	−0.37	−0.43	−0.43
C^3	0.10	0.10	0.10	0.10	0.14	0.14	0.17	0.17
C^4	−0.33	−0.33	−0.34	−0.34	−0.36	−0.36	−0.39	−0.39
C^5	0.30	0.31	0.30	0.31	0.33	0.33	0.30	0.30
C^6	−0.37	−0.38	−0.40	−0.40	−0.37	−0.35	−0.33	−0.33
C^9	−0.23	−0.23	−0.22	−0.22	−0.24	−0.25	−0.20	−0.20
C^{10}	−0.15	−0.13	−0.09	−0.09	−0.07	−0.07	−0.12	−0.09
C^{13}	−0.05	−0.05	−0.09	−0.05	−0.06	−0.05	−0.08	−0.05
C^{14}	0.34	0.35	0.38	0.35	0.35	0.34	0.38	0.35
C^{19}	0.00	−0.01	0.00	0.00	−0.08	−0.08	0.01	0.02
C^{22}	−0.33	−0.34	−0.39	−0.34	−0.34	−0.34	−0.38	−0.34
O^1	−0.88	−0.89	−0.93	−0.93	−0.91	−0.91	−0.93	−0.92
O^3	−0.87	−0.87	−0.89	−0.89	−0.80	−0.80	−0.76	−0.76
O^5	−0.89	−0.88	−0.93	−0.93	−0.82	−0.82	−0.80	−0.80

(Table 4.7) contd.....

Parameter	C1$_4$	C2$_4$	C1$_{3A}$	C2$_{3A}$	C1$_3$	C2$_3$	T1$_3$	T2$_3$
O^{11}	−0.81	−0.81	−0.82	−0.81	−0.92	−0.93	−0.92	−0.92
O^{19}	−0.87	−0.87	−0.80	−0.80	−0.85	−0.85	−0.91	−0.91
Ca	1.65	1.65	1.69	1.69	1.67	1.68	1.69	1.69

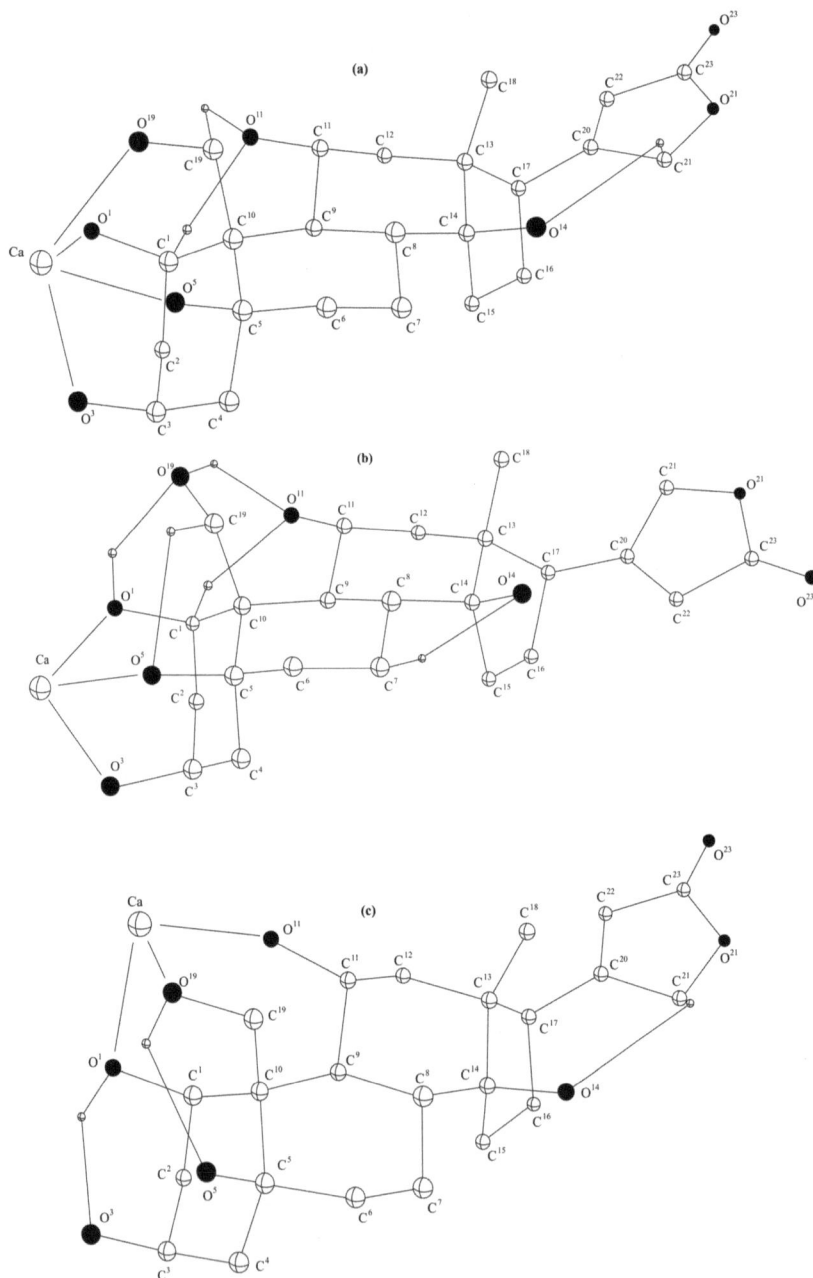

Fig. 4.11 contd.....

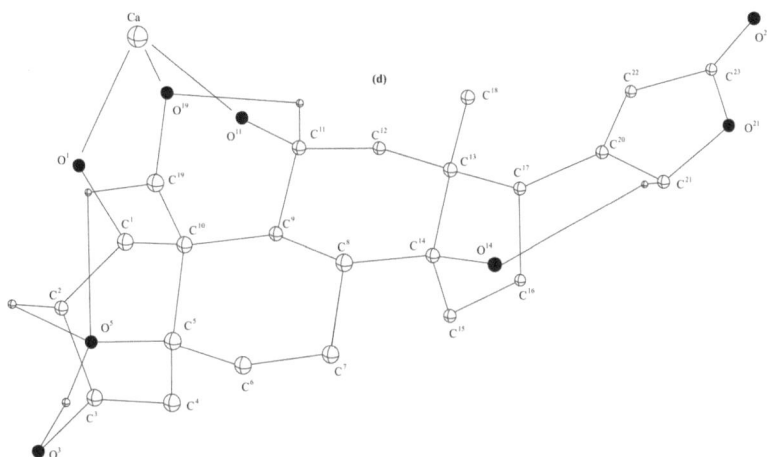

Fig. (4.11). Steric structures of selected ouabagenin–Ca^{2+} conformations, accepted atom numbering, and intramolecular hydrogen bonding patterns:
a – Conformation $C2_4$;
b – Conformation $C1_{3A}$;
c – Conformation $C2_3$;
d – Conformation $T2_3$.
Only some of hydrogen atoms are shown. Hydrogen bonds are presented as dotted lines; coordination Ca–O bonds, as arrows.

Comparison of the data given in Tables **4.1** and **4.7** made it possible to estimate the effect of chelation on the ligand structure. The C—O bonds with oxygen atoms involved in Ca^{2+} chelation are elongated by 0.02–0.09 Å. Several endocyclic C—C bonds in rings A and B change within 0.02 Å. In conformation family T, C^2—C^3 and C^5—C^{10} bonds become longer by 0.03 Å. Fairly noticeable changes are observed for endocyclic bonds in ring D: C^{14}—C^{15} and C^{15}—C^{16} bonds are extended by 0.03-0.05 and 0.01-0.02 Å, respectively, whereas C^{13}—C^{17} and C^{16}—C^{17} bonds shorten by 0.03-0.05 and 0.01-0.02 Å, respectively. These variations in interatomic distances are accompanied by changes of the bond angle values, especially of those at the coordinating oxygen atoms and at C^5 and C^{10} atoms. $T1_{3A}$ and $T1_3$ conformations are also characterized by substantial variations of the bond angles involving C^{14} atom, and $C^{16}C^{17}C^{13}$ angle decreases by ~2°. In conformation family 3, variations of several endocyclic bond angles in rings B and C are appreciable. $C^{12}C^{11}O^{11}$ angles in $C1_{3A}$, $C2_{3A}$, $C1_4$, and $C2_4$ conformations increase by ~4°, while $C^9C^{11}O^{11}$ angles decrease by roughly the same value.

Torsion angle values are also affected upon Ca^{2+} chelation. Ring A becomes flatter: τ_{av} in examined conformations of ouabagenin–Ca^{2+} complexes is smaller by 5–7° than in free ouabagenin. The τ_{av} value for ring B decreases by ~3° in $C1_3$ and $C2_3$ conformations and increases by ~1.5° in $T1_3$ and $T2_3$ conformations; τ_{av} for ring C remains almost the same as in ouabagenin (except for $T1_3$ and $T2_3$ conformations, where it decreases by ~2°). Ca^{2+} chelation induces considerable conformational changes in ring D. In free ouabagenin, ring D has the 14β-*envelope* conformation in conformation family 2 and the *half-chair* conformation in conformation family 1 (see above). In conformation family 2 of ouabagenin–Ca^{2+} chelates, the 14β-*envelope* conformation becomes noticeably distorted. Moreover, ring D in $C2_3$ conformation adopts the *half-chair* conformation with C^{14} and C^{15} atoms located on opposite sides of the plane formed by C^{13}, C^{16}, and C^{17}. The same shape of ring D was detected in $C1_4$ conformation. Rings D in $C1_{3A}$ and $T1_3$ conformations may be regarded as distorted 13α-*envelopes;* and in $C1_3$, as a distorted 15α-*envelope* (Table **4.7**). Ca^{2+} chelation results in substantial changes of the values of torsion angles determining mutual orientation of rings A and B in all conformations, except for $C1_{3A}$ and $C2_{3A}$. Orientation of rings B and C and of rings C and D with respect to each other, as well as of the lactone ring with respect to the steroid core changes rather slightly.

The most appreciable variations in the Mulliken atomic charges upon chelation of Ca^{2+} are observed for the oxygen atoms directly involved in Ca^{2+} coordination: the charges on these atoms become more negative by 0.06–0.14 a.u. The charge on O^{23} atom decreases by 0.04 a.u. in the absolute value. Essential redistribution of electron density occurs in ring A. In conformation family C, the charges on C^1 and C^3 atoms decrease by 0.03-0.08 a.u., the charge on C^4 atom becomes less negative by 0.01-0.04 a.u., and the charge on C^{10} atom increases in the absolute value by 0.01-0.06 a.u. In conformation family T, the charge on C^1 atom increases by 0.03 a.u., and the charges on C^2 and C^4 atoms increase by 0.04-0.07 a.u. in the absolute value. All conformations of ouabagenin–Ca^{2+} complexes are characterized by more negative values of the charges on C^6, C^{12}, and C^{15} atoms (by 0.01-0.07 a.u. as compared to the free ligand).

The described variations in the geometry and electronic structure of ouabagenin upon Ca^{2+} chelation are largely determined by weakening of C—O bonds which include O^1, O^3, O^5, O^{11}, and O^{19} atoms directly involved in coordination of Ca^{2+}. As a result, the values of corresponding bond angles and the charges on these oxygen atoms and carbon atoms bonded to them also change. The most pronounced changes in structural and electronic parameters are observed in ring A, since only this ring contains hydroxy groups donating electron density upon Ca^{2+} chelation in all examined conformations of ouabagenin–Ca^{2+}. Conformational reorganization of ring D may be attributed to changes in intramolecular hydrogen bonding patterns, the parameters of which are presented in Table **4.8**. Insofar as oxygen atoms of ouabagenin are mostly involved in coordination of Ca^{2+}, there are fewer hydrogen bonds in ouabagenin–Ca^{2+} complexes than in free ouabagenin.

Several hydrogen $O\cdots H-C$ bonds appear: conformations of family 1 are stabilized by $O^{14}\cdots H-C^7$ bonds; those of family 2, by $O^{14}\cdots H-C^{21}$ bonds. Hydrogen bonds which involve O^{14} atom in the free ligand are considerably weaker; therefore, strengthening of these hydrogen bonds upon Ca^{2+} chelation appreciably influences geometry and electronic parameters of ring D. Analysis of intramolecular hydrogen bonding made it possible to explain other less significant structural changes of ouabagenin resulting from chelation of Ca^{2+}. For instance, the values of $C^9C^{11}O^{11}$ and $C^{12}C^{11}O^{11}$ bond angles in $C1_{3A}$, $C2_{3A}$, $C1_4$, and $C2_4$ conformations are determined by the presence of $O^{11}\cdots H-C^1$ hydrogen bond. Change of the bond angle values at C^{14} atom in $C1_{3A}$ and $T1_3$ is related to the fact that $O^{14}\cdots H-C^7$ hydrogen bonds in these conformations are stronger than those in $C1_3$ and $C1_4$ conformations.

Table 4.8. Calculated parameters of intramolecular hydrogen bonds[a] **present in examined conformations of ouabagenin–Ca^{2+} chelate complex: R_1 (Å), $O\cdots H$ distance; R_2 (Å), $O\cdots X$ distance.**

$C1_4$			$C2_4$			$C1_3$			$C2_3$		
$O\cdots X$ Atoms	R_1	R_2	$O\cdots X$ Atoms	R_1	R_2	$O\cdots X$ Atoms	R_1	R_2	$O\cdots X$ Atoms	R_1	R_2
$O^{11}\cdots C^1$	1.98	2.83	$O^{11}\cdots C^1$	1.97	2.83	$O^3\cdots O^1$	1.86	2.66	$O^3\cdots O^1$	1.87	2.67
$O^{11}\cdots C^{19}$	2.34	3.08	$O^{11}\cdots C^{19}$	2.37	3.10	$O^5\cdots O^{19}$	2.05	2.68	$O^5\cdots O^{19}$	2.07	2.69
$O^{14}\cdots C^7$	2.51	2.87	$O^{14}\cdots C^{21}$	2.71	3.30	$O^{14}\cdots C^7$	2.48	2.85	$O^{14}\cdots C^{21}$	2.91	3.54
$C1_{3A}$			$C2_{3A}$			$T1_3$			$T2_3$		
$O\cdots X$ Atoms	R_1	R_2	$O\cdots X$ Atoms	R_1	R_2	$O\cdots X$ Atoms	R_1	R_2	$O\cdots X$ Atoms	R_1	R_2
$O^{19}\cdots O^1$	1.87	2.61	$O^{19}\cdots O^1$	1.88	2.61	$O^5\cdots C^{19}$	2.25	2.58	$O^5\cdots C^{19}$	2.25	2.58
$O^{11}\cdots C^1$	2.06	2.88	$O^{11}\cdots C^1$	2.04	2.86	$O^3\cdots O^5$	2.29	2.84	$O^3\cdots O^5$	2.26	2.84
$O^{11}\cdots O^{19}$	2.10	2.88	$O^{11}\cdots O^{19}$	2.15	2.92	$O^{19}\cdots C^{11}$	2.37	2.95	$O^{19}\cdots C^{11}$	2.36	2.95
$O^5\cdots C^{19}$	2.34	2.87	$O^5\cdots C^{19}$	2.34	2.87	$O^{14}\cdots C^7$	2.38	2.78	$O^5\cdots C^2$	2.40	2.78
$O^{14}\cdots C^7$	2.37	2.73	$O^{14}\cdots C^{21}$	2.69	3.29	$O^{11}\cdots C^1$	2.53	3.00	$O^{14}\cdots C^{21}$	2.71	3.32

[a]Intramolecular hydrogen bonds are considered in $O\cdots H-X$ form, where X = O or C.

Distances $r(Ca-O)$ in optimized structures of ouabagenin–Ca^{2+} complexes are almost similar for conformation families 1 and 2. In $C1_4$ conformation: $r(Ca-O^1)$ = 2.32 Å, $r(Ca-O^3)$ = 2.42 Å, $r(Ca-O^5)$ = 2.31 Å, $r(Ca-O^{19})$ = 2.39 Å; in $C1_{3A}$ conformation: $r(Ca-O^1)$ = 2.29 Å, $r(Ca-O^3)$ = 2.31 Å, $r(Ca-O^5)$ = 2.33 Å; in $C1_3$ conformation: $r(Ca-O^1)$ = 2.30 Å, $r(Ca-O^{11})$ = 2.50 Å, $r(Ca-O^{19})$ = 2.33 Å; in $T1_3$ conformation: $r(Ca-O^1)$ = 2.29 Å, $r(Ca-O^{11})$ = 2.31 Å, $r(Ca-O^{19})$ = 2.32 Å. Increase of $r(Ca-O^1)$ and $r(Ca-O^{19})$ by 0.02–0.07 Å in conformation family 4 is determined by electrostatic repulsion between the oxygen atoms and rigidity of the steroid core which does not allow four oxygen atoms to simultaneously approach the cation more closely.

A relatively small number of steric and electronic parameters in examined ouabagenin–Ca^{2+} conformations remain fairly constant. These parameters for $C2_4$ conformation are collected in Table **4.9**, and they describe mostly the structure of the lactone ring. Also, the majority of endocyclic C—C bond lengths in rings A, B, and C change insignificantly. Correspondingly, chelation-induced conformational reorganization of the steroid core involves mainly variation of bond and torsion angles.

Table 4.9. Selected calculated (RHF/6-31G*) bond lengths, bond angles and atomic charges in conformation $C2_4$ of ouabagenin–Ca^{2+} chelate complex.

Bond Length, r, Å									
C^1—C^2	1.526	C^7—C^8	1.525	C^{12}—C^{13}	1.543	C^{20}—C^{22}	1.321	C^{21}—O^{21}	1.405
C^3—C^4	1.519	C^8—C^9	1.555	C^{13}—C^{18}	1.536	C^{22}—C^{23}	1.482	C^{23}—O^{21}	1.344
C^4—C^5	1.537	C^8—C^{14}	1.556	C^{15}—C^{16}	1.555	C^{14}—O^{14}	1.410	C^{23}—O^{23}	1.178
C^5—C^6	1.529	C^9—C^{11}	1.553	C^{17}—C^{20}	1.510				
C^6—C^7	1.517	C^{11}—C^{12}	1.527	C^{20}—C^{21}	1.516				
Bond Angle, ω, deg									
$C^5C^6C^7$	113.4	$C^{12}C^{13}C^{14}$	108.1	$C^{13}C^{17}C^{20}$	116.7	$C^{17}C^{13}C^{18}$	113.6	$C^{22}C^{23}O^{21}$	107.1
$C^6C^5C^{10}$	113.1	$C^{12}C^{13}C^{17}$	107.7	$C^{14}C^{13}C^{17}$	104.2	$C^{20}C^{22}C^{23}$	109.3	$C^{22}C^{23}O^{23}$	129.1
$C^6C^7C^8$	110.2	$C^{12}C^{13}C^{18}$	109.2	$C^{14}C^{13}C^{18}$	113.7	$C^{21}C^{20}C^{22}$	107.6	$C^{21}O^{21}C^{23}$	111.3
$C^7C^8C^{14}$	111.1	$C^{13}C^{14}C^8$	113.5	$C^{15}C^{16}C^{17}$	106.5	$C^{15}C^{14}O^{14}$	109.2	$O^{21}C^{23}O^{23}$	123.8
$C^8C^9C^{11}$	105.4	$C^{13}C^{14}C^{15}$	103.6	$C^{16}C^{17}C^{20}$	113.2	$C^{20}C^{21}O^{21}$	104.8		
Atomic Charge, q, a.u.									
C^7	−0.36	C^{12}	−0.37	C^{17}	−0.20	C^{21}	0.01	O^{21}	−0.62
C^8	−0.18	C^{15}	−0.37	C^{18}	−0.49	C^{23}	0.82	O^{23}	−0.53
C^{11}	0.20	C^{16}	−0.33	C^{20}	0.01	O^{14}	−0.78		

Quantum-chemical calculations allowed us to reveal some differences in the structures of ouabain and ouabagenin calcium chelates. Ring A in ouabagenin–Ca^{2+} complexes can have both *chair* and *twist* conformations, whereas this ring adopts the *boat* conformation in ouabain–Ca^{2+} complexes. The most illustrative way to compare the geometry parameters of ouabagenin and ouabain calcium chelates is direct superposition of the corresponding conformations, as shown in Fig. (**4.12**). Only a few steric differences between the structures of ouabagenin–Ca^{2+} and ouabain–Ca^{2+} are observed. It is important that the spatial positions of coordinated cations almost coincide. Therefore, the critical difference in the mechanisms of $Na_V1.8$ channels modulation by ouabain and ouabagenin cannot be attributed to different geometric parameters of their calcium complexes, and it is most probably accounted for by the lack of the rhamnosyl residue in ouabagenin.

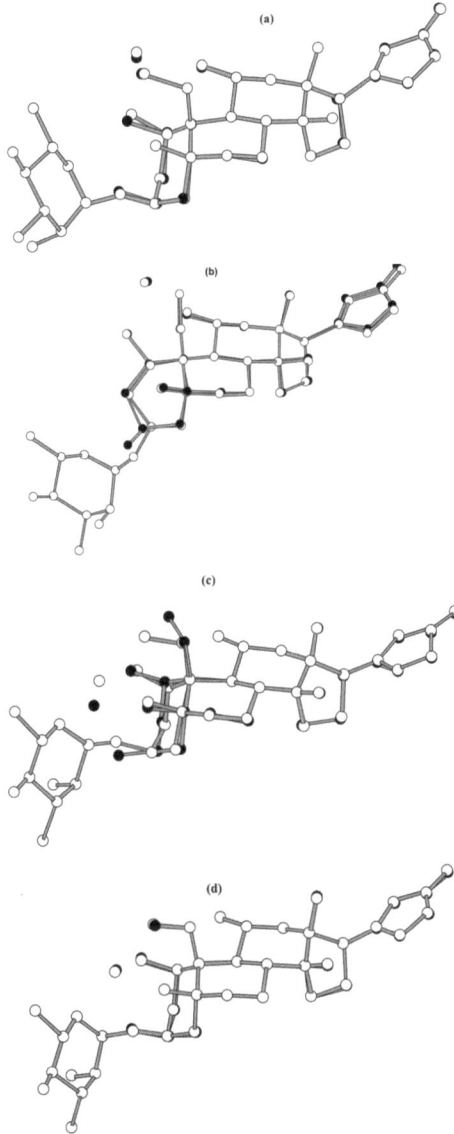

Fig. (4.12). Superposition of steric structures of ouabain–Ca^{2+} and ouabagenin–Ca^{2+}:
a – Conformations $C2_3$;
b – Conformation $B2_3$ (ouabain–Ca^{2+}) and conformation $T2_3$ (ouabagenin–Ca^{2+});
c – Conformation $C2_5$ (ouabain–Ca^{2+}) and conformation $C2_{3A}$ (ouabagenin–Ca^{2+});
d – Conformation $C2_5$ (ouabain–Ca^{2+}) and conformation $C2_4$ (ouabagenin–Ca^{2+}).
Light circles correspond to atoms in ouabain–Ca^{2+}; dark circles, ouabagenin–Ca^{2+}. Hydrogen atoms are not shown.

Marinobufagenin

Marinobufagenin (3β,5β-dihydroxy-14,15β-epoxy-5β,14β-bufa-20,22-dienolide) is another cardiotonic steroid, which was demonstrated above by the patch-clamp method to exhibit $Na_V1.8$ channel-modulating effect similar to that of ouabagenin. However, marinobufagenin belongs to the class of bufadienolides distinguished by a six-membered lactone ring, unlike cardenolides ouabain and ouabagenin containing a five-membered lactone ring. No published data is available on steric structure of marinobufagenin. X-ray diffraction analysis was carried out for cinobufotalin (16β-acetoxy-3β,5β-dihydroxy-14,15β-epoxy-5β,14β-bufa-20,22-dienolide) [61] differing from marinobufagenin solely by the presence of 16β-acetoxy group (Fig. **4.13**). Experimental parameters of this molecule were used to test the adequacy of our quantum-chemical calculations of marinobufagenin.

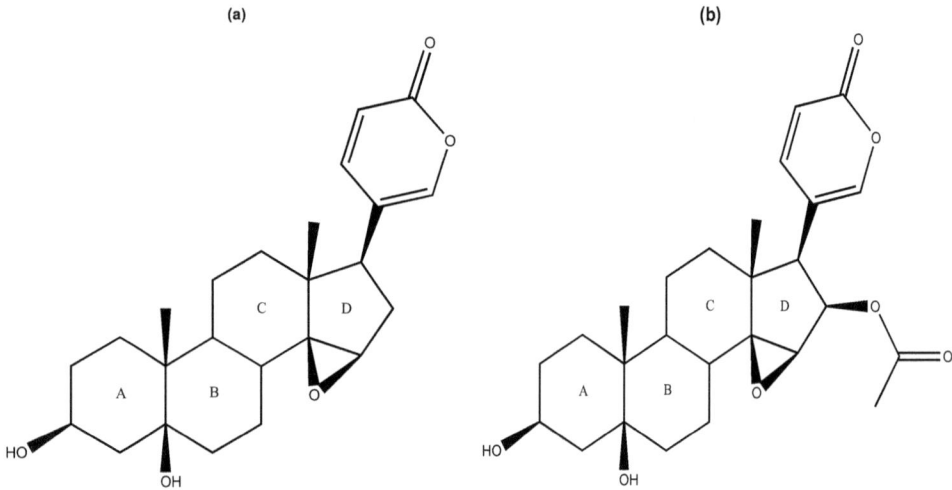

Fig. (4.13). Stereochemical formulae of marinobufagenin (a) and cinobufotalin (b).

As the steroid core structure and the lactone ring location are the same in marinobufagenin, ouabain, and ouabagenin, four possible conformations of marinobufagenin were considered and their geometry parameters were fully optimized *ab initio* by the restricted Hartree–Fock method with the 6-31G* basis set [54] using GAMESS program [55]. Steric structures of two marinobufagenin conformations are presented in Fig. (**4.14**). Indices C and T designate the conformation of ring A (*chair* and *twist*, respectively), indices 1 and 2 relate to orientation of the lactone ring with respect to the steroid core ($\tau(C^{16}C^{17}C^{20}C^{21})$ values are about 125° and -42°, respectively). According to our calculations, C1 conformation has the lowest energy, while those of C2, T1 and T2 conformations are higher by ~2, ~6 and ~8 kcal·mol⁻¹, respectively. This is in conformity with the

XRD data on cinobufotalin, which adopts C1 conformation in the crystal lattice [61].

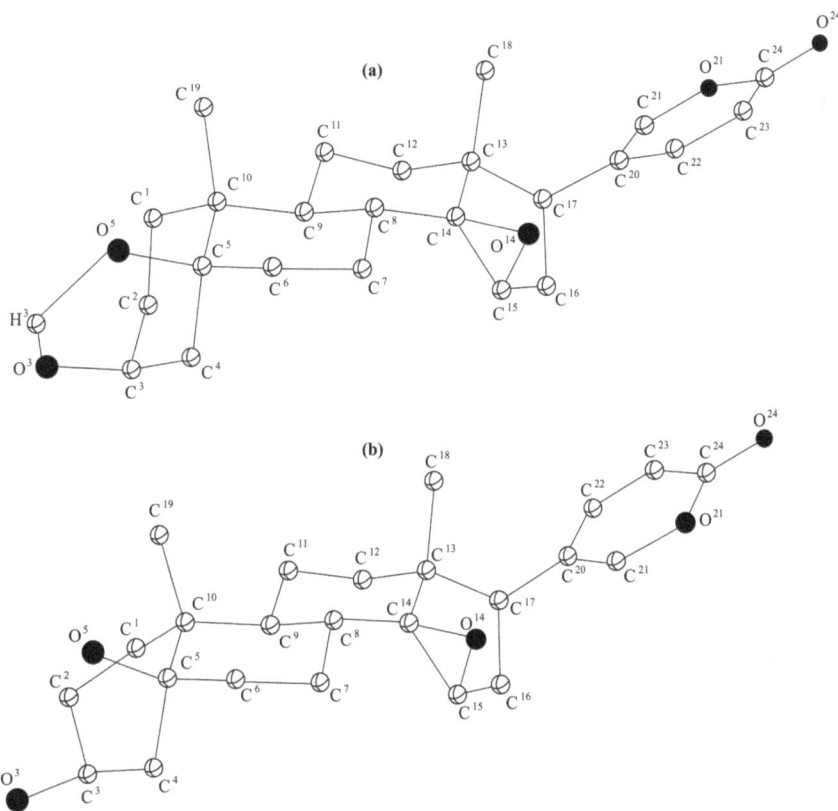

Fig. (4.14). Steric structures of selected marinobufagenin conformations and accepted atom numbering:
a – Conformation C1;
b – Conformation T2.
Hydrogen atoms except H^3 are not shown. Hydrogen bond is presented by the dotted line.

Selected bond lengths, bond angles, torsion angles, and atomic charges in all studied marinobufagenin conformations are presented in Table **4.10**. The calculated parameters are in a fair agreement with the values obtained experimentally for cinobufotalin [61]. The bond lengths differ mostly by less than 0.015 Å; the bond angles, by less than 1.5°. A considerable difference is observed only in the values of $C^2C^3O^3$, $C^4C^3O^3$, $C^4C^5O^5$, and $C^6C^5O^5$ bond angles describing the positions of O^3 and O^5 atoms, and in C—O bond lengths, which in the marinobufagenin molecule are by 0.03-0.05 Å shorter than the corresponding bonds in cinobufotalin.

Table 4.10. Selected calculated (RHF/6-31G*) bond lengths, bond and torsion angles, and atomic charges in examined conformations of marinobufagenin.

Parameter	C1	C2	T1	T2	Parameter	C1	C2	T1	T2
Bond Length, r, Å									
C^1—C^{10}	1.550	1.550	1.564	1.564	C^3—O^3	1.404	1.403	1.418	1.418
C^2—C^3	1.528	1.528	1.537	1.537	C^5—O^5	1.424	1.424	1.418	1.418
Bond Angle, ω, deg									
$C^1C^{10}C^5$	108.1	108.1	109.4	109.4	$C^5C^{10}C^9$	109.2	109.2	110.7	110.7
$C^1C^{10}C^9$	111.9	111.9	107.1	107.1	$C^6C^5O^5$	108.2	108.2	105.5	105.5
$C^1C^{10}C^{19}$	106.2	106.3	108.8	108.9	$C^6C^7C^8$	111.5	111.5	109.9	110.0
$C^2C^3C^4$	110.3	110.3	111.9	111.9	$C^7C^8C^9$	111.9	112.1	110.4	110.5
$C^3C^4C^5$	114.6	114.6	111.9	111.8	$C^8C^9C^{10}$	112.6	112.7	114.7	114.7
$C^4C^3O^3$	112.5	112.5	106.6	106.6	$C^9C^{10}C^{19}$	110.7	110.8	111.6	111.6
$C^4C^5C^6$	110.4	110.4	112.3	112.3	$C^{10}C^5O^5$	110.1	110.2	108.7	108.6
$C^4C^5C^{10}$	112.2	112.2	109.1	109.1	$C^{10}C^9C^{11}$	113.9	113.9	112.9	112.9
$C^4C^5O^5$	103.7	103.7	108.5	108.5	$C^{17}C^{20}C^{21}$	120.0	125.1	120.0	125.1
$C^5C^6C^7$	113.6	113.6	114.8	114.9	$C^{17}C^{20}C^{22}$	124.4	119.6	124.4	119.6
Torsion Angle, τ, deg									
$C^1C^2C^3C^4$	-53.0	-53.1	29.4	29.3	$C^6C^7C^8C^9$	-52.9	-52.6	-57.4	-57.2
$C^1C^{10}C^5C^4$	50.7	50.7	37.1	37.1	$C^7C^6C^5C^{10}$	-54.8	-54.9	-51.3	-51.3
$C^2C^1C^{10}C^5$	-53.9	-54.0	24.0	24.0	$C^{12}C^{13}C^{14}C^{15}$	102.3	100.8	102.7	101.0
$C^2C^3C^4C^5$	53.2	53.2	32.2	32.4	$C^{12}C^{13}C^{14}O^{14}$	165.9	164.5	166.2	164.6
$C^3C^2C^1C^{10}$	56.6	56.6	-60.4	-60.3	$C^{13}C^{17}C^{16}C^{15}$	-19.6	-22.8	-19.5	-22.8
$C^3C^4C^5C^{10}$	-53.3	-53.3	-68.5	-68.6	$C^{14}C^{13}C^{17}C^{16}$	19.7	22.6	19.4	22.5
$C^4C^5C^{10}C^9$	-71.3	-71.2	-80.7	-80.7	$C^{14}C^{15}C^{16}C^{17}$	12.5	14.8	12.6	14.8
$C^5C^6C^7C^8$	53.8	53.7	57.4	57.3	$C^{15}C^{14}C^{13}C^{17}$	-12.5	-14.2	-12.2	-14.1
$C^5C^{10}C^9C^8$	-53.9	-53.7	-48.1	-48.1	$C^{16}C^{17}C^{20}C^{21}$	125.3	-41.6	124.8	-41.6
$C^6C^5C^{10}C^9$	53.4	53.5	44.6	44.7	$C^{16}C^{17}C^{20}C^{22}$	-52.5	137.2	-53.0	137.1
Atomic Charge, q, a.u.									
C^1	-0.33	-0.33	-0.35	-0.34	C^{19}	-0.52	-0.51	-0.48	-0.48
C^2	-0.33	-0.33	-0.35	-0.35	C^{20}	-0.08	-0.10	-0.07	-0.10
C^3	-0.18	-0.18	-0.17	-0.17	C^{21}	0.16	0.19	0.16	0.19
C^4	-0.35	-0.35	-0.38	-0.38	O^3	-0.79	-0.79	-0.77	-0.77
C^6	-0.34	-0.34	-0.31	-0.31	O^5	-0.80	-0.80	-0.81	-0.81
C^{10}	-0.03	-0.03	-0.01	-0.01					

Newman projections of marinobufagenin and cinobufotalin in C1 conformations are presented in Fig. (**4.15**). The relative positions of C and D rings in these molecules are almost identical (projections along C^{14}—C^{13} bonds, Fig. **4.15.b**). The relative positions of A and B rings, as well as the positions of the lactone ring with respect to the steroid core slightly differ (projections along C^5—C^{10} and C^{17}—C^{20} bonds, Figs. **4.15.a** and **4.15.c**, respectively), which is due to a number of factors: neglection of the crystal lattice effects in our calculations of marinobufagenin, presence of a bulky acetoxy group in cinobufotalin that affects the lactone ring orientation, and difference in intra- and intermolecular hydrogen bonding patterns. Intermolecular hydrogen bond between O^3 atom and one of oxygen atoms of the acetoxy group detected in the crystal lattice of cinobufotalin [61] is apparently missing in marinobufagenin. Intramolecular hydrogen bond $O^3 \cdots H$—O^5 [$r(O\cdots O) = 2.77$ Å, $r(H\cdots O) = 1.70$ Å)] present in cinobufotalin [61] is replaced with $O^5 \cdots H$—O^3 bond [$r(O\cdots O) = 2.79$ Å, $r(H\cdots O) = 2.01$ Å)] in marinobufagenin.

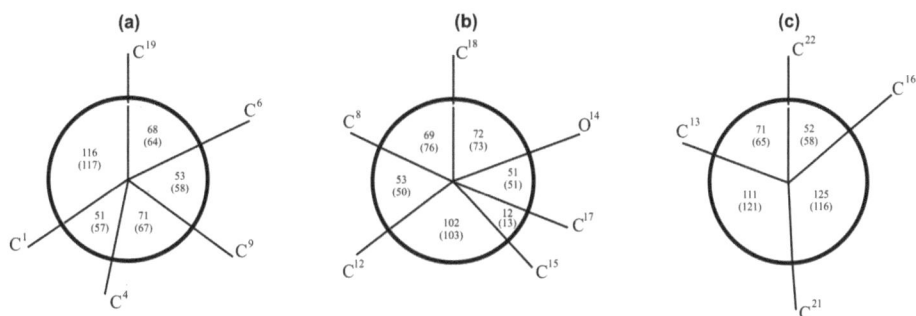

Fig. (4.15). Newman projections of marinobufagenin and cinobufotalin along C^5—C^{10} (a), C^{14}—C^{13} (b) and C^{17}—C^{20} (c) bonds. Data of [61] are given in parentheses.

Since the calculated geometry for C1 conformation of marinobufagenin is well consistent with the XRD data, it may be presumed that the optimized parameters of all other conformations are also sufficiently realistic. Rings B and C are in the *chair* conformation, ring A adopts either the *chair* or the *twist* conformation, and ring D is in the 17β-*envelope* conformation, same as in cinobufotalin [61], due to the epoxy cycle. The lactone ring is nearly planar: the absolute values of endocyclic torsion angles do not exceed 0.7°. The geometric and electronic parameters of marinobufagenin which remain fairly constant in all studied conformations are given in Table **4.11** for C1 conformation.

Analyzing the data presented in Tables **4.10** and **4.11**, the following conclusions regarding the difference in structural and electronic parameters of marinobufagenin conformations can be made. The bond distances practically do not vary, except for two C—C bonds in ring A, C^3—O^3 and C^5—O^5 bonds. The bond angles change mainly within ring A, to a lesser extent within ring B. The torsion angles in ring A vary essentially, and the differences are also detected in the shape of rings B and D. The lactone ring and ring C adopt the same shape in all conformations. Changes of the atomic charge values are observed on atoms of rings A and B, on lactone ring atoms, as well as on O^3 and O^5 atoms.

Table 4.11. Selected calculated (RHF/6-31G*) bond lengths, bond angles, and atomic charges in conformation C1 of marinobufagenin.

Bond Length, r, Å									
C^1—C^2	1.528	C^7—C^8	1.531	C^{11}—C^{12}	1.531	C^{14}—C^{15}	1.453	C^{20}—C^{22}	1.457
C^3—C^4	1.531	C^8—C^9	1.549	C^{12}—C^{13}	1.543	C^{15}—O^{14}	1.414	C^{21}—O^{21}	1.342
C^4—C^5	1.541	C^8—C^{14}	1.520	C^{13}—C^{14}	1.527	C^{15}—C^{16}	1.506	C^{22}—C^{23}	1.335
C^5—C^6	1.533	C^9—C^{10}	1.569	C^{13}—C^{17}	1.578	C^{16}—C^{17}	1.555	C^{23}—C^{24}	1.459
C^5—C^{10}	1.568	C^9—C^{11}	1.540	C^{13}—C^{18}	1.533	C^{17}—C^{20}	1.518	C^{24}—O^{21}	1.358
C^6—C^7	1.526	C^{10}—C^{19}	1.544	C^{14}—O^{14}	1.414	C^{20}—C^{21}	1.330	C^{24}—O^{24}	1.358

Bond Angle, ω, deg									
$C^1C^2C^3$	111.4	$C^8C^{14}C^{15}$	126.5	$C^{12}C^{13}C^{18}$	110.1	$C^{14}C^{15}O^{14}$	59.1	$C^{20}C^{21}O^{21}$	124.4
$C^2C^1C^{10}$	115.0	$C^8C^{14}O^{14}$	116.9	$C^{13}C^{14}C^{15}$	109.2	$C^{14}O^{14}C^{15}$	61.8	$C^{21}C^{20}C^{22}$	115.6
$C^2C^3O^3$	111.4	$C^9C^8C^{14}$	110.2	$C^{13}C^{14}O^{14}$	117.6	$C^{15}C^{14}O^{14}$	59.1	$C^{21}O^{21}C^{24}$	123.1
$C^5C^{10}C^{19}$	110.6	$C^9C^{11}C^{12}$	112.3	$C^{13}C^{17}C^{16}$	105.5	$C^{15}C^{16}C^{17}$	105.4	$C^{22}C^{23}C^{24}$	121.1
$C^6C^5C^{10}$	111.8	$C^{11}C^{12}C^{13}$	113.9	$C^{13}C^{17}C^{20}$	112.6	$C^{16}C^{15}O^{14}$	113.3	$C^{23}C^{24}O^{21}$	115.0
$C^7C^8C^{14}$	112.6	$C^{12}C^{13}C^{14}$	106.3	$C^{14}C^{13}C^{17}$	104.9	$C^{16}C^{17}C^{20}$	113.2	$C^{23}C^{24}O^{24}$	126.2
$C^8C^9C^{11}$	110.1	$C^{12}C^{13}C^{14}$	106.3	$C^{14}C^{13}C^{18}$	112.5	$C^{17}C^{13}C^{18}$	113.3	$O^{21}C^{24}O^{24}$	118.8
$C^8C^{14}C^{13}$	118.5	$C^{12}C^{13}C^{17}$	108.5	$C^{14}C^{15}C^{16}$	110.9	$C^{20}C^{22}C^{23}$	120.8		

Torsion Angle, τ, deg									
$C^7C^8C^9C^{10}$	54.5	$C^8C^9C^{11}C^{12}$	54.5	$C^9C^{11}C^{12}C^{13}$	-57.3	$C^{11}C^9C^8C^{14}$	-53.0	$C^{13}C^{14}C^{15}C^{16}$	0.1
$C^7C^8C^9C^{11}$	-177.1	$C^8C^{14}C^{13}C^{12}$	-52.6	$C^9C^8C^{14}C^{13}$	53.9	$C^{11}C^{12}C^{13}C^{14}$	52.3		

Atomic Charge, q, a.u.									
C^5	0.32	C^{11}	-0.33	C^{15}	0.07	C^{22}	-0.12	O^{14}	-0.63
C^7	-0.33	C^{12}	-0.32	C^{16}	-0.34	C^{23}	-0.36	O^{21}	-0.62
C^8	-0.18	C^{13}	-0.02	C^{17}	-0.22	C^{24}	0.81	O^{24}	-0.56
C^9	-0.18	C^{14}	0.31	C^{18}	-0.48				

We have established by the patch-clamp method that ouabain, ouabagenin, and marinobufagenin upon extracellular application to sensory neurons of rat spinal ganglia decrease the effective charge (Z_{eff}) of activation gating device of $Na_V1.8$

channels (Table **4.12**), which indicates the potential ability of these cardiotonic steroids to exhibit analgesic properties. However, investigating combined action of marinobufagenin (ouabagenin) and ouabain, it was demonstrated that additional application of ouabain at a concentration of 200 μM, sufficient to block the pumping function of Na^+,K^+-ATPase and thus interrupt the signal transduction from this enzyme to $Na_V 1.8$ channels, weakly affected the decrease of Z_{eff}. It may be therefore concluded that at very low "endogenous" concentration the mechanism of marinobufagenin and ouabagenin action is different from that of ouabain.

Table 4.12. Effective charges (Z_{eff}) of activation gating device of $Na_V 1.8$ channels after extracellular application of cardiotonic steroids.

Substance	Concentration, nM	Z_{eff}, a.u.	Number of Experiments
Control		6.6 ± 0.3	24
Ouabain	1	4.8 ± 0.3^a	15
Ouabagenin	100	4.8 ± 0.3^a	9
Ouabagenin with preliminary application of ouabain (200 μM)	100	5.0 ± 0.3^a	20
Marinobufagenin	10	4.7 ± 0.4^a	15
Marinobufagenin with preliminary application of ouabain (200 μM)	10	4.9 ± 0.2^a	9

^{a} difference between experimental and control data is statistically significant

Presented in Fig. (**4.16**) superpositions of steric structures of marinobufagenin and ouabagenin (ouabain) in C2 conformations, which are the lowest in energy for ouabain and ouabagenin, demonstrate that the structural difference between the three molecules is relatively small, despite the absence in marinobufagenin of fairly intricate intramolecular hydrogen bonding patterns existing in ouabagenin and ouabain. Hence, to account for the ability of ouabain to modulate both pumping and signal-transducing functions of Na^+,K^+-ATPase, we suggest that two distinct molecular forms of this steroid are involved: free ouabain blocks the pump, while ouabain–Ca^{2+} chelate complex activates the signal transduction. Direct verification of this hypothesis in a calcium-free extracellular solution by the patch-clamp method is associated with technical difficulties.

However, the results of additional experiments carried out by the organotypic tissue culture method indicate that combined application of ouabain and EGTA, a selective Ca^{2+} chelator, almost completely blocks the inhibiting action of the former on tissue culture growth [62]. Moreover, the ouabagenin effect on cell growth and proliferation in five different tissue types was shown not to be tissue-specific, in contrast to the ouabain effect in the same tissues, and it was due to modulation of the Na^+,K^+-ATPase pumping function [62]. Ouabain and ouabagenin molecules possess a sufficient number of oxygen atoms capable to participate in Ca^{2+} chelation, and superposition of ouabain–Ca^{2+} and ouabagenin–Ca^{2+} spatial structures did not reveal any essential differences in their geometry (Fig. **4.12**), whereas the location of oxygen atoms in marinobufagenin does not permit formation of stable Ca^{2+} chelate complexes in 1:1 stoichiometry. Thus, an additional factor accounting for the difference between the mechanisms of ouabain and ouabagenin/marinobufagenin action should exist. This factor is the rhamnosyl ring in ouabain, which contains four oxygen atoms capable to form intermolecular hydrogen bonds upon ouabain binding to the transducing site of Na^+,K^+-ATPase.

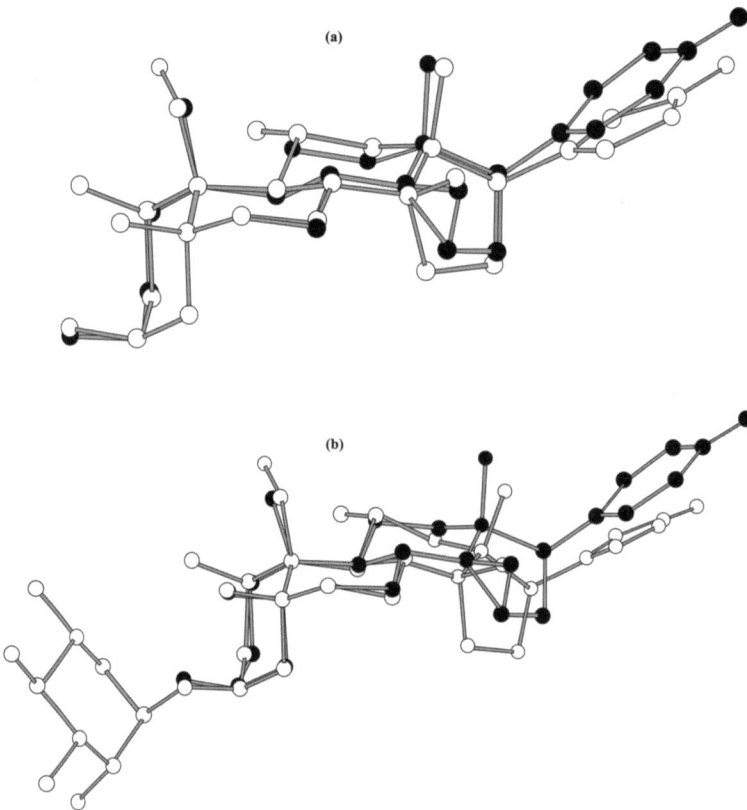

Fig. (4.16). Superposition of steric structures of cardiotonic steroids in C2 conformations
a – marinobufagenin and ouabagenin;
b – marinobufagenin and ouabain.
Light circles correspond to atoms in ouabain and ouabagenin; dark circles, marinobufagenin. Hydrogen atoms are not shown.

CONCLUSION

Summarizing the results described in the current chapter, the mechanisms of interaction of studied cardiotonic steroids with Na^+,K^+-ATPase can have the following interpretation. The pumping function of Na^+,K^+-ATPase is inhibited by free ouabain, ouabagenin, and marinobufagenin molecules at micromolar concentrations. The signal-transducing function of this enzyme is modulated exclusively by ouabain in the form of its calcium chelate complex in nanomolar range of concentrations, and the significant energy contribution to formation of the ternary ouabain–Ca^{2+}–Na^+,K^+-ATPase complex is provided not only by ion-ionic bond(s) between the chelated calcium ion and nucleophilic functional group(s) of the Na^+,K^+-ATPase transducing site but also by intermolecular hydrogen bonds involving oxygen atoms of the rhamnosyl residue. This fragment is missing in ouabagenin, and the energy of interaction of ouabagenin–Ca^{2+} chelate complex with Na^+,K^+-ATPase is insufficient to activate the signal-transducing function of the enzyme, despite the ability of ouabagenin to form such complexes and structural similarity of ouabagenin and ouabain calcium chelates. The decrease of Z_{eff} of $Na_V 1.8$ channel activation gating device induced by application of ouabagenin or marinobufagenin at nanomolar concentrations (Table **4.12**) results from activation of the "modulated receptor" mechanism, *i.e.*, these molecules bind directly to the aminoacid sequence of the channel without involvement of Ca^{2+}.

It must be stressed that the current views on ouabain function should be reconsidered. First, utilization of a variety of accurate quantitative methods and criteria made it possible to demonstrate that ouabain at subnanomolar endogenous concentrations should bind to a specific target in the sensory neuron membrane. This target site is located directly on Na^+,K^+-ATPase and it recognizes ouabain only in the form of its calcium chelate complex. Activation of this site switches on the transducer-coupled mechanism of modulation of $Na_V 1.8$ channels, which play a crucial role in nociception. Second, at higher concentrations ouabain performs its well-known function of the sodium pump inhibition. We suggest that this process is triggered by binding of free ouabain to another Na^+,K^+-ATPase site. According to our estimate, the K_d values of these two processes differ by approximately 4 orders of magnitude.

Activation of the first process by an endogenous substance is presumed to be a physiologically adequate adaptive response of the organism to stress or damage. This response results in pain relief and, likely, in triggering regeneration and reparative processes. The sodium pump inhibition can probably also produce a curative effect, but only insofar as chronic pathological state is caused by excessive activity of excitable tissue.

Our results based on theoretical quantum-chemical calculations and experimental patch-clamp results demonstrate the dual effect of ouabain and provide an alternative explanation of the mechanism of Na^+,K^+-ATPase control by this agent. Two distinct attacking molecules (ouabain and its calcium chelate complex) bind to two distinct sites of the sodium pump, thus modulating two distinct functions of the enzyme: pumping and non-pumping (signal-transducing).

CONFLICT OF INTEREST

The authors confirm that they have no conflict of interest to declare for this publication.

ACKNOWLEDGEMENTS

Declared none.

REFERENCES

[1]   Buckalew VM. Endogenous digitalis-like factors: an overview of the history. Front Endocrinol (Lausanne) 2015; 6(49): 49.
        [PMID: 25918512]

[2]   Lichtstein D. Na^+,K^+-ATPase and heart excitability. In: Sideman S, Beyar R, Eds. Molecular and subcellular cardiology: effects of structure and function. New York: Plenum Press 1995; pp. 23-30.
        [http://dx.doi.org/10.1007/978-1-4615-1893-8_3]

[3]   Reddy BA. Digitalis therapy in patients with congestive heart failure. Int J Pharm Sci Rev Res 2010; 3(2): 90-5.

[4]   Blaustein MP, Zhang J, Chen L, Hamilton BP. How does salt retention raise blood pressure? Am J Physiol Regul Integr Comp Physiol 2006; 290(3): R514-23.
        [http://dx.doi.org/10.1152/ajpregu.00819.2005] [PMID: 16467498]

[5]   Weidemann H. Na/K-ATPase, endogenous digitalis-like compounds and cancer development - a hypothesis. Front Biosci 2005; 10: 2165-76.
        [http://dx.doi.org/10.2741/1688] [PMID: 15970485]

[6]   Goldstein I, Levy T, Galili D, *et al.* Involvement of Na($^+$), K($^+$)-ATPase and endogenous digitalis-like compounds in depressive disorders. Biol Psychiatry 2006; 60(5): 491-9.
        [http://dx.doi.org/10.1016/j.biopsych.2005.12.021] [PMID: 16712803]

[7]   Goldstein I, Lax E, Gispan-Herman I, *et al.* Neutralization of endogenous digitalis-like compounds alters catecholamines metabolism in the brain and elicits anti-depressive behavior. Eur Neuropsychopharmacol 2012; 22(1): 72-9.
        [http://dx.doi.org/10.1016/j.euroneuro.2011.05.007] [PMID: 21700431]

[8]   Schoner W, Scheiner-Bobis G. Endogenous cardiac glycosides: hormones using the sodium pump as signal transducer. Semin Nephrol 2005; 25(5): 343-51.
        [http://dx.doi.org/10.1016/j.semnephrol.2005.03.010] [PMID: 16139690]

[9]   Vakkuri O, Arnason SS, Pouta A, Vuolteenaho O, Leppäluoto J. Radioimmunoassay of plasma ouabain in healthy and pregnant individuals. J Endocrinol 2000; 165(3): 669-77.
        [http://dx.doi.org/10.1677/joe.0.1650669] [PMID: 10828851]

[10]   Yamada K, Goto A, Hui C, *et al.* Role of ouabainlike compound in rats with reduced renal mass-saline hypertension. Am J Physiol 1994; 266(4 Pt 2): H1357-62.

[PMID: 8184913]

[11]   Aydemir-Koksoy A, Abramowitz J, Allen JC. Ouabain-induced signaling and vascular smooth muscle cell proliferation. J Biol Chem 2001; 276(49): 46605-11.
[http://dx.doi.org/10.1074/jbc.M106178200] [PMID: 11579090]

[12]   Chueh SC, Guh JH, Chen J, Lai MK, Teng CM. Dual effects of ouabain on the regulation of proliferation and apoptosis in human prostatic smooth muscle cells. J Urol 2001; 166(1): 347-53.
[http://dx.doi.org/10.1016/S0022-5347(05)66157-5] [PMID: 11435898]

[13]   Abramowitz J, Dai C, Hirschi KK, *et al*. Ouabain- and marinobufagenin-induced proliferation of human umbilical vein smooth muscle cells and a rat vascular smooth muscle cell line, A7r5. Circulation 2003; 108(24): 3048-53.
[http://dx.doi.org/10.1161/01.CIR.0000101919.00548.86] [PMID: 14638550]

[14]   Orlov SN, Thorin-Trescases N, Pchejetski D, *et al*. Na+/K+ pump and endothelial cell survival: [Na+]i/[K+]i-independent necrosis triggered by ouabain, and protection against apoptosis mediated by elevation of [Na+]i. Pflügers Arch 2004; 448(3): 335-45.
[http://dx.doi.org/10.1007/s00424-004-1262-9] [PMID: 15069561]

[15]   Watabe M, Nakajo S, Yoshida T, Kuroiwa Y, Nakaya K. Treatment of U937 cells with bufalin induces the translocation of casein kinase 2 and modulates the activity of topoisomerase II prior to the induction of apoptosis. Cell Growth Differ 1997; 8(8): 871-9.
[PMID: 9269896]

[16]   Terness P, Navolan D, Dufter C, Kopp B, Opelz G. The T-cell suppressive effect of bufadienolides: structural requirements for their immunoregulatory activity. Int Immunopharmacol 2001; 1(1): 119-34.
[http://dx.doi.org/10.1016/S0162-3109(00)00264-2] [PMID: 11367509]

[17]   Hamlyn JM, Ringel R, Schaeffer J, *et al*. A circulating inhibitor of $(Na^+ + K^+)$ATPase associated with essential hypertension. Nature 1982; 300(5893): 650-2.
[http://dx.doi.org/10.1038/300650a0] [PMID: 6292738]

[18]   Hamlyn JM, Blaustein MP, Bova S, *et al*. Identification and characterization of a ouabain-like compound from human plasma. Proc Natl Acad Sci USA 1991; 88(14): 6259-63.
[http://dx.doi.org/10.1073/pnas.88.14.6259] [PMID: 1648735]

[19]   Hamlyn JM, Hamilton BP, Manunta P. Endogenous ouabain, sodium balance and blood pressure: a review and a hypothesis. J Hypertens 1996; 14(2): 151-67.
[http://dx.doi.org/10.1097/00004872-199602000-00002] [PMID: 8728291]

[20]   Dobretsov M, Stimers JR. Neuronal function and alpha3 isoform of the Na/K-ATPase. Front Biosci 2005; 10: 2373-96.
[http://dx.doi.org/10.2741/1704] [PMID: 15970502]

[21]   Crambert G, Hasler U, Beggah AT, *et al*. Transport and pharmacological properties of nine different human Na, K-ATPase isozymes. J Biol Chem 2000; 275(3): 1976-86.
[http://dx.doi.org/10.1074/jbc.275.3.1976] [PMID: 10636900]

[22]   Müller-Ehmsen J, Juvvadi P, Thompson CB, *et al*. Ouabain and substrate affinities of human Na(+)-K(+)-ATPase α(1)β(1), α(2)β(1), and α(3)β(1) when expressed separately in yeast cells. Am J Physiol Cell Physiol 2001; 281(4): C1355-64.
[PMID: 11546674]

[23]   Wang J, Velotta JB, McDonough AA, Farley RA. All human Na(+)-K(+)-ATPase α-subunit isoforms have a similar affinity for cardiac glycosides. Am J Physiol Cell Physiol 2001; 281(4): C1336-43.
[PMID: 11546672]

[24]   Sweadner KJ. Enzymatic properties of separated isozymes of the Na,K-ATPase. Substrate affinities, kinetic cooperativity, and ion transport stoichiometry. J Biol Chem 1985; 260(21): 11508-13.
[PMID: 2995339]

[25] Inoue N, Matsui H, Tsukui H, Hatanaka H. The appearance of a highly digitalis-sensitive isoform of Na+,K+-ATPase during maturation *in vitro* of primary cultured rat cerebral neurons. J Biochem 1988; 104(3): 349-54.
[http://dx.doi.org/10.1093/oxfordjournals.jbchem.a122472] [PMID: 2853703]

[26] Blanco G, Mercer RW. Isozymes of the Na-K-ATPase: heterogeneity in structure, diversity in function. Am J Physiol 1998; 275(5 Pt 2): F633-50.
[PMID: 9815123]

[27] Dvela M, Rosen H, Feldmann T, Nesher M, Lichtstein D. Diverse biological responses to different cardiotonic steroids. Pathophysiology 2007; 14(3-4): 159-66.
[http://dx.doi.org/10.1016/j.pathophys.2007.09.011] [PMID: 17964766]

[28] Hamada K, Matsuura H, Sanada M, *et al.* Properties of the Na+/K+ pump current in small neurons from adult rat dorsal root ganglia. Br J Pharmacol 2003; 138(8): 1517-27.
[http://dx.doi.org/10.1038/sj.bjp.0705170] [PMID: 12721107]

[29] Krylov BV, Derbenev AV, Podzorova SA, Liudyno MI, Kuzmin AV, Izvarina NL. [Morphine decreases the voltage sensitivity of the slow sodium channels]. Ross Fiziol Zh Im I M Sechenova 1999; 85(2): 225-36.
[PMID: 10389179]

[30] Krylov BV, Derbenev AV, Podzorova SA, Lyudyno MI, Kuzmin AV, Izvarina NL. Morphine decreases the voltage sensitivity of slow sodium channels. Neurosci Behav Physiol 2000; 30(4): 431-9.
[http://dx.doi.org/10.1007/BF02463098] [PMID: 10981947]

[31] Xie Z. Ouabain interaction with cardiac Na/K-ATPase reveals that the enzyme can act as a pump and as a signal transducer. Cell Mol Biol (Noisy-le-grand) 2001; 47(2): 383-90.
[PMID: 11357899]

[32] Li Z, Xie Z. The Na/K-ATPase/Src complex and cardiotonic steroid-activated protein kinase cascades. Pflügers Arch 2009; 457(3): 635-44.
[http://dx.doi.org/10.1007/s00424-008-0470-0] [PMID: 18283487]

[33] Bagrov AY, Fedorova OV, Dmitrieva RI, *et al.* Characterization of a urinary bufodienolide Na+,K+-ATPase inhibitor in patients after acute myocardial infarction. Hypertension 1998; 31(5): 1097-103.
[http://dx.doi.org/10.1161/01.HYP.31.5.1097] [PMID: 9576120]

[34] Fedorova LV, Raju V, El-Okdi N, *et al.* The cardiotonic steroid hormone marinobufagenin induces renal fibrosis: implication of epithelial-to-mesenchymal transition. Am J Physiol Renal Physiol 2009; 296(4): F922-34.
[http://dx.doi.org/10.1152/ajprenal.90605.2008] [PMID: 19176701]

[35] Fedorova OV, Doris PA, Bagrov AY. Endogenous marinobufagenin-like factor in acute plasma volume expansion. Clin Exp Hypertens 1998; 20(5-6): 581-91.
[http://dx.doi.org/10.3109/10641969809053236] [PMID: 9682914]

[36] Gallice PM, Kovacic HN, Brunet PJ, Berland YF, Crevat AD. A non ouabain-like inhibitor of the sodium pump in uremic plasma ultrafiltrates and urine from healthy subjects. Clin Chim Acta 1998; 273(2): 149-60.
[http://dx.doi.org/10.1016/S0009-8981(98)00032-1] [PMID: 9657345]

[37] Gonick HC, Ding Y, Vaziri ND, Bagrov AY, Fedorova OV. Simultaneous measurement of marinobufagenin, ouabain, and hypertension-associated protein in various disease states. Clin Exp Hypertens 1998; 20(5-6): 617-27.
[http://dx.doi.org/10.3109/10641969809053240] [PMID: 9682918]

[38] Harwood S, Mullen AM, McMahon AC, Dawnay A. Plasma OLC is elevated in mild experimental uremia but is not associated with hypertension. Am J Hypertens 2001; 14(11 Pt 1): 1112-5.
[http://dx.doi.org/10.1016/S0895-7061(01)02219-1] [PMID: 11724209]

[39]   Kennedy DJ, Vetteth S, Periyasamy SM, *et al.* Central role for the cardiotonic steroid marinobufagenin in the pathogenesis of experimental uremic cardiomyopathy. Hypertension 2006; 47(3): 488-95.
[http://dx.doi.org/10.1161/01.HYP.0000202594.82271.92] [PMID: 16446397]

[40]   Komiyama Y, Dong XH, Nishimura N, *et al.* A novel endogenous digitalis, telocinobufagin, exhibits elevated plasma levels in patients with terminal renal failure. Clin Biochem 2005; 38(1): 36-45.
[http://dx.doi.org/10.1016/j.clinbiochem.2004.08.005] [PMID: 15607315]

[41]   Li S, Liu G, Jia J, *et al.* Therapeutic monitoring of serum digoxin for patients with heart failure using a rapid LC-MS/MS method. Clin Biochem 2010; 43(3): 307-13.
[http://dx.doi.org/10.1016/j.clinbiochem.2009.09.025] [PMID: 19833118]

[42]   Manunta P, Stella P, Rivera R, *et al.* Left ventricular mass, stroke volume, and ouabain-like factor in essential hypertension. Hypertension 1999; 34(3): 450-6.
[http://dx.doi.org/10.1161/01.HYP.34.3.450] [PMID: 10489392]

[43]   Periyasamy SM, Chen J, Cooney D, *et al.* Effects of uremic serum on isolated cardiac myocyte calcium cycling and contractile function. Kidney Int 2001; 60(6): 2367-76.
[http://dx.doi.org/10.1046/j.1523-1755.2001.00053.x] [PMID: 11737612]

[44]   Liu J, Tian J, Haas M, Shapiro JI, Askari A, Xie Z. Ouabain interaction with cardiac Na^+/K^+-ATPase initiates signal cascades independent of changes in intracellular Na^+ and Ca^{2+} concentrations. J Biol Chem 2000; 275(36): 27838-44.
[PMID: 10874029]

[45]   Pierre SV, Xie Z. The Na,K-ATPase receptor complex: its organization and membership. Cell Biochem Biophys 2006; 46(3): 303-16.
[http://dx.doi.org/10.1385/CBB:46:3:303] [PMID: 17272855]

[46]   Pavlovic D. The role of cardiotonic steroids in the pathogenesis of cardiomyopathy in chronic kidney disease. Nephron Clin Pract 2014; 128(1-2): 11-21.
[http://dx.doi.org/10.1159/000363301] [PMID: 25341357]

[47]   Lopatina EV, Yachnev IL, Penniyaynen VA, *et al.* Modulation of signal-transducing function of neuronal membrane Na+,K+-ATPase by endogenous ouabain and low-power infrared radiation leads to pain relief. Med Chem 2012; 8(1): 33-9.
[http://dx.doi.org/10.2174/157340612799278531] [PMID: 22420548]

[48]   Yachnev IL, Plakhova VB, Podzorova SA, Shelykh TN, Rogachevsky IV, Krylov BV. Mechanism of pain relief by low-power infrared irradiation: ATP is an IR-target molecule in nociceptive neurons. Med Chem 2012; 8(1): 14-21.
[http://dx.doi.org/10.2174/157340612799278595] [PMID: 22420546]

[49]   Messerschmidt A. Ouabain, $C_{29}H_{44}O$, 2-BHZO. Cryst Struct Commun 1980; 9(3): 1185-93.

[50]   Go K, Kartha G. Ouabain diethanol, $C_{29}H_{44}O_{12}.2C_2H_5OH$. Cryst Struct Commun 1981; 10(4): 1329-34.

[51]   Go K, Kartha G. Structure of ouabagenin methanol solvate, $C_{23}H_{34}O_8.CH_3OH$. Acta Cryst 1983; C39(3): 376-8.

[52]   McIntyre DD, Germann MW, Vogel HJ. Conformational analysis and complete assignment of the proton and carbon NMR spectra of ouabain and ouabagenin. Can J Chem 1990; 68(8): 1263-70.
[http://dx.doi.org/10.1139/v90-195]

[53]   Bohl M, Sussmilch R. Calculations on molecular structure and electrostatic potentials of cardiotonic steroids. Eur J Med Chem 1986; 21(3): 193-8.
[PMID: 6492795]

[54]   Hariharan P, Pople J. The influence of polarization functions on molecular orbital hydrogenation energies. Theor Chim Acta 1973; 28(3): 213-22.
[http://dx.doi.org/10.1007/BF00533485]

[55] Schmidt M, Baldridge K, Boatz J, *et al.* General atomic and molecular electronic structure system. J Comput Chem 1993; 14(11): 1347-63.
[http://dx.doi.org/10.1002/jcc.540141112]

[56] Kawamura A, Abrell LM, Maggiali F, *et al.* Biological implication of conformational flexibility in ouabain: observations with two ouabain phosphate isomers. Biochemistry 2001; 40(19): 5835-44.
[http://dx.doi.org/10.1021/bi0101751] [PMID: 11341849]

[57] Rohrer DC, Duax WL, Fullerton DS. Structures of modified cardenolides. II. (20R)-3β-Hydroxy-22-methylene-5β-card-14-enoline. Acta Cryst 1976; B32(10): 2893-5.
[http://dx.doi.org/10.1107/S056774087600914X]

[58] Kawamura KI. Microscale stereochemical analyses of sphingoid bases and brassinosteroids. II. On the structure and physiology of endogenous ouabain. PhD dissertation. New York: Columbia University 1999.

[59] Melero CP, Medarde M, San Feliciano A. A short review on cardiotonic steroids and their aminoguanidine analogues. Molecules 2000; 5(1): 51-81.
[http://dx.doi.org/10.3390/50100051]

[60] Handschuh S, Goldfuss B, Chen J, Gasteiger J, Houk KN. Steroid binding by antibodies and artificial receptors: exploration of theoretical methods to determine the origins of binding affinities and specificities. J Comput Aided Mol Des 2000; 14(7): 611-29.
[http://dx.doi.org/10.1023/A:1008188322239] [PMID: 11008884]

[61] Kálmán A, Fülöp V, Argay G, *et al.* Structures of two bufadienolides: bufotalin (3β, 14-dihydroxy-16β-acetoxy-5β, 14β-bufa-20, 22-dienolide) and cinobufotalin (3β, 5β-dihydroxy-14, 15β-epoxy-16β -acetoxy-5β, 14β-bufa-20, 22-dienolide). Acta Cryst 1988; C44(9): 1634-8.

[62] Penniiaĭnen VA, Kipenko AV, Lopatina EV, Krylov BV. [The influence of ouabagenin on the growth and proliferation of cells in the organotypical culture]. Ross Fiziol Zh Im I M Sechenova 2014; 100(11): 1303-9.
[PMID: 25665409]

CHAPTER 5

Concluding Remarks

Abstract: Molecular mechanisms of the nociceptive information control in primary sensory neuron are described based on our investigation of the membrane signaling cascade (opioid-like receptor $\rightarrow$ Na$^+$,K$^+$-ATPase $\rightarrow$ Na$_V$1.8 channel). Summarizing the data presented in this volume it is possible to conclude that modulation of Na$_V$1.8 channels responsible for the coding of noxious signals can be carried out due to two novel targeting mechanisms. The first of these is the activation of opioid-like receptors; the second is the activation of the Na$^+$,K$^+$-ATPase signal-transducing function. Development of a novel class of analgesics that trigger these mechanisms should lead in the near future to successful solution of the problem of chronic pain relief.

Keywords: Analgesic, Modulated receptor, Na$_V$1.8 channels, Na$^+$,K$^+$-ATPase, nociception, Opioid-like receptor, Signal transducer.

It is known that pain is unpleasant but necessary. It signals of danger, preventing us from harming ourselves, and alerts on possible damage to our bodies. Too much pain is crippling and can make everyday living an agony. That is why pain and suicide are related. Even "good" pain can turn bad, when the pain caused by an injury persists after the damage has healed. Chronic pain dramatically reduces the quality of life for millions of people. There is no doubt that any steps to develop potent and safe analgesics are of major importance. Unfortunately, no analgesics in the arsenal of practical medicine satisfy these two criteria at the same time. However, there is always hope that other opportunities to fight pain are hidden within the human body. Even the smallest practical result in finding them is very important, because endogenous mechanisms of pain relief should have no negative side effects. In our opinion, to elucidate them it is necessary to link physiology, which is the basis of medical science, with calculational chemistry that makes it possible to describe physiological events on the detailed molecular level.

The basic physiological principles should be applied to analyze the fundamental mechanisms of nociception. Ivan Pavlov was the first who revealed a strong coupling between internal inhibition processes and antagonistic nervous process of excitation [1]. Intensity of sensory signals in the peripheral nervous system is

Boris V. Krylov, Ilia V. Rogachevskii, Tatiana N. Shelykh, Vera B. Plakhova

simply coded by frequency of nerve impulses. This principle discovered by Edgar Adrian [2] is widely used by us in this book. We also rely on the assumption of Vernon Mountcastle, who formulated it as the linear operator principle [3, 4]. We take this principle into consideration when we quantitatively describe the processes of receptor- or transducer-coupled modulation of $Na_V 1.8$ channels. In other words, we postulate that the process of ligand-receptor binding that occurs in neighboring protein molecule linearly (or monotonically) influences the effective charge value of $Na_V 1.8$ channels activation gating system.

We believe that the novel mechanism of $Na_V 1.8$ channels modulation in nociceptive neuron (Fig. **1.7**) will open a new approach to solve the problem of chronic pain. Fig. (**1.7**) indicates the presence of three separate molecular targets. Each of them can interact with its "own" agonists and antagonists, some of which should be endogenous. It can be thus argued that the physiological effects of these interactions should result in the control of nociceptive signals. In accordance with our approach, antinociceptive response of sensory neuron can be obtained through activation of three different molecular mechanisms triggered by three different targets: opioid-like receptor, Na^+,K^+-ATPase as a signal transducer, and $Na_V 1.8$ channel.

OPIOID-LIKE RECEPTOR-COUPLED MECHANISM OF $NA_V 1.8$ CHANNELS MODULATION

Several unexpected manifestations of morphine action were presented in Chapter **1**. We propose a completely new additional explanation of powerful analgesic effect produced by this substance. It is assumed that the opioid-like receptor-coupled mechanism is also responsible for the analgesic effect of morphine. Of course, the agent runs the well-studied opioidergic system, activation of which leads to pain relief. One terrible disadvantage property intrinsic to opioid receptor agonists, however, does not allow morphine to become an ideal analgesic. This disadvantage is the appearance of multiple negative side effects as a result of its systematic application. It is tempting to speculate that the cause of the adverse side effects of the agent at the molecular level is its ability to activate G proteins coupled to classic opioid receptors. We have found a fundamentally different mechanism of morphine action, the role of signal transducer in which is performed by Na^+,K^+-ATPase of nociceptive neuron [5]. Now it becomes clear that the analgesic effect of morphine is of dual nature: it activates both classic opioid receptors and opioid-like receptors physiologically described in the present book. Our results suggest that a selective agonist of novel opioid-like receptors will be free of negative side effects, since in this case the transducing function would be performed by Na^+,K^+-ATPase and not by G proteins. As it was shown in Chapters **1-3**, activation of the Na^+,K^+-ATPase transducing function is a marker of involvement of opioid-like receptors in modulation of $Na_V1.8$ channels. Identification of the selective agonist of opioid-like receptors which differs essentially from morphine both structurally and physiologically is the important result of our work. This agent is comenic acid, which, unlike morphine, binds selectively only to opioid-like receptors, thus resulting in modulation of $Na_V1.8$ channels responsible for coding of nociceptive signals.

Combined application of the patch-clamp method and quantum-chemical calculations made it possible to clarify the difference between morphine and comenic acid in their receptor-coupled ability to modulate $Na_V1.8$ channels. The latter substance, being of a significantly smaller molecular volume than morphine, specifically activates only opioid-like receptors due to its remarkable property: it can chelate calcium ions from the surrounding physiological medium. Our study on the effects of gamma-pyrone and gamma-pyridone derivatives presented in Chapters **2** and **3** allows to describe the probable characteristics of opioid-like receptor binding pocket and get an insight on molecular structure of the endogenous agonist of these receptors, which is not yet identified.

Four of six studied molecules (substances A, B, E, and F) displayed the ability to modulate $Na_V1.8$ channels, while the other two (substances C and D) were inactive. Our results made it possible to establish that the active substances should bind to the opioid-like receptor being in the form of calcium salt of calcium chelate complex and to consequently formulate the structural criteria determining the possibility for formation of ligand-receptor complexes between gamma-pyridones or gamma-pyrones and the opioid-like receptor: (1) in position 5 of the heterocycle should be present a hydroxyl or methoxy group which is capable, in combination with the carbonyl group in position 4, to chelate Ca^{2+} cation; (2) the second Ca^{2+} cation serves as the counterion at the deprotonated carboxyl or hydroxymethyl group in position 2 of the heterocycle; (3) intercationic distance $r(Ca^{2+}\cdots Ca^{2+})$ may range from 9.4 to 10.0 Å; and (4) Ca^{2+} cations should occupy specific positions with respect to the heterocycle. The major contribution to the energy of ligand-receptor binding of gamma-pyrones and gamma-pyridones is provided by strong ion-ionic interactions between bound calcium cations of the ligand and negatively charged aspartate residues of the opioid-like receptor. It is also found that the nature of the ring heteroatom may influence the ability of ligands to bind to the opioid-like receptor. Substance F, a structural gamma-pyridone analog of inactive gamma-pyrone D, exhibits $Na_V1.8$ channel-modulating effect due to the presence of intramolecular hydrogen bond between the heterocycle nitrogen atom and the oxygen atom of the hydroxymethyl group in position 2 of the pyridone ring, which fixates this substituent in the conformation appropriate for ligand-receptor binding. Several observations are made regarding the structure of opioid-like receptor binding pocket: it is, most probably, rather small and has a slit-like shape, so the planar molecules of comenic acid and other active gamma-pyrones and gamma-pyridones should be accommodated rather snugly. The morphine molecule is likely to occupy the whole volume of the binding pocket, forming a very stable ligand-receptor complex, since endogenous μ-opioid agonists endomorphin-1 and endomorphin-2 are not demonstrated to exhibit any $Na_v1.8$ channel-modulating activity [6], which indicates that these molecules are too large to be accommodated in opioid-like receptor binding pocket.

It can be predicted that an as yet undiscovered endogenous molecule, which activates the opioid-like receptor at physiologically appropriate conditions in nanomolar or subnanomolar concentration range, should exist in the human organism. The suggestion that comenic acid is similar in its characteristics to this hypothetical molecule can account for the success of preclinical trials of novel analgesic Anoceptin® which contains the agent as the drug substance. Also, the results of Phase I clinical studies have shown that Anoceptin® is absolutely safe for administration in humans [7]. The clinical data obtained draw attention to several very important facts. First, none of the subjects experienced any adverse side effects inherent to activation of the opioidergic system after application of comenic acid. Fortunately, no opioid-related disorders were registered. Also, comenic acid never induced the miosis, a characteristic reaction of humans on administration of opioid drugs. Second, the results of pharmacokinetic analysis demonstrate that comenic acid almost completely disappears from the blood stream within minutes [7]. At the same time its analgesic effect lasts for hours. It is well known that morphine produces its analgesic effect only when present in the blood serum in sufficient concentration for hours [8]. We can conclude that binding of comenic acid to opioid-like receptors safely and effectively modulates the nociceptive system. This modulation process induces strong analgesic effect, substantially distinct from the morphine action. It means that there is an additional mechanism of the nociceptive information processing in the human brain, which is studied rather poorly so far. Further investigation of this mechanism will result in creation of fundamentally new analgesics, by means of which the problem of chronic pain relief should be resolved. We believe that the development of Anoceptin® is the first step in this direction [9].

TRANSDUCER-COUPLED MECHANISM OF $NA_v1.8$ CHANNELS MODULATION

Fig. (**1.17**) indicates that Na^+,K^+-ATPase should play a key role in modulation of nociceptive signals. The results described in Chapter **4** show that the specific inhibitor of the sodium pump, ouabain, should have dual targeting on one and the same Na^+,K^+-ATPase molecule. Combined application of the patch-clamp method and quantum-chemical calculations makes it possible to explain that only free ouabain molecule, without chelated Ca^{2+}, inhibits the Na^+,K^+-ATPase pumping function by the well-known mechanism of interaction with its low affinity site. Modulation of $Na_V1.8$ channels is governed by another mechanism which involves Na^+,K^+-ATPase as the signal transducer. Quantum-chemical calculations demonstrate that in adequate physiological conditions ouabain should take the form of a stable chelate complex with calcium ion in 1:1 stoichiometry. It is ouabain–Ca^{2+} complex that activates the high affinity site of Na^+,K^+-ATPase responsible for activation of its transducing function. So, Na^+,K^+-ATPase performing its additional non-pumping function in sensory neuron, very likely, serves as the receptor for endogenous ouabain. Activation of signal transduction results in $Na_V1.8$ channel modulation. It means that the nociceptive system is under specific control of a well-known endogenous substance, ouabain. It may be predicted that fine tuning of nociceptive signals is organism-specific because of high diversity and heterogeneity of Na^+,K^+-ATPase isoforms [10].

On the other hand, ouabain is widely used in medicine for two hundred years due to its ability to inhibit the pumping function of Na^+,K^+-ATPase [11]. Its application results in the increase of heart contraction and cardiac output in patients with heart failure [12, 13]. Clinical application of ouabain is restricted by its extremely high toxicity, since the capacity of sarcoplasmic reticulum Ca^{2+} storage is exceeded following excessive ouabain-induced Na^+,K^+-ATPase inhibition and Na^+ accumulation, thus leading to generation of delayed after depolarizations and arrhythmias [13]. Our data obtained on embryonic nerve tissue indicate that ouabain also inhibits neurite growth with K_d of 0.1 nM [14].

Our approach allows to shed light on the solution to the problem of conflicting effects of ouabain. It is free calcium ions that perform a particularly important regulating function. Usually, the extracellular concentration of free Ca^{2+} is extremely low in regular physiological conditions. Ouabain, as an endogenous factor secreted by the adrenal glands in humans, is present in blood serum at subnanomolar concentrations. Interaction of ouabain with Ca^{2+} in the extracellular medium results in formation of ouabain–Ca^{2+} chelate complex. Our data indicate that only this complex is capable of activating the signal-transducing function of Na^+,K^+-ATPase, which is the mechanism of modulation of $Na_V1.8$ channels and control of the nociceptive information processing. As discussed above, morphine exhibits the dual effect, because it activates different targets: opioid receptors and opioid-like receptors. The dual effect of ouabain is essentially different: we have shown that it interacts with two distinct Na^+,K^+-ATPase binding sites. One of them is the well-known low affinity site of "free" ouabain binding, which results in inhibition of the pumping function of the enzyme. The other site characterized by high affinity to ouabain can be activated only by ouabain–Ca^{2+} chelate complex, which indicates that calcium ions are thus involved in the regulation of nociceptive signals, of course, in a complex with endogenous ouabain. Ouabain–Ca^{2+} chelate complex and "free" ouabain are two different molecules, but they share a rather similar chemical structure. As a consequence, the two binding sites should have many common structural motifs. It seems rather reasonable to assume that they almost overlap. In other words, it is the energy of additional ion-ionic bond(s) between the chelated Ca^{2+} and nucleophilic functional group(s) of Na^+,K^+-ATPase that determines the ability of ouabain to activate the signal-transducing function of the sodium pump by forming this specific interaction between the ligand and the enzyme.

Antinociceptive effect of ouabain was never described before. It was demonstrated to be triggered by activation of the Na^+,K^+-ATPase signal-transducing function [14]. Na^+,K^+-ATPase-coupled antinociceptive effect has also been observed independently: we have shown that low-power infrared radiation with the wavelength of 10.6 μm also can specifically control the signal-transducing function of Na^+,K^+-ATPase, which leads to modulation of $Na_V1.8$ channels [15]. Thus, the results of basic physiological investigations can find a wide application in physiotherapy. It should be emphasized that this method of peripheral chronic pain relief is completely safe even for the long-term systematic use.

MODULATED RECEPTOR MECHANISM OF $NA_V1.8$ CHANNELS CONTROL

The receptor-coupled mechanism of modulation of nociceptive signals by gamma-pyrone derivatives described in the present volume is fundamentally new, and it is protected by a number of our patents [16, 17]. At the same time, a completely different approach to modulation of $Na_V1.8$ channels has been proposed by other authors [18, 19], who have found several potent and effective blockers of $Na_V1.8$ channels. However, these substances interact directly with the $Na_V1.8$ channel protein by the modulated receptor mechanism, which is radically different from the mechanism of action of gamma-pyrones, as the signal transduction is not required in this case. On the contrary, successful application of comenic acid as the drug substance is based on the mechanisms of receptor-coupled molecular recognition and amplification of the triggered signal by Na^+,K^+-ATPase as the transducer. This novel membrane signaling cascade physiologically adequately controls $Na_V1.8$ channels in a very effective and specific manner, which is supported by the results of the first phase of clinical trials of new analgesic Anoceptin®: no adverse side effects were revealed.

Physiological role of the modulated receptor mechanism in nociception remains to be elucidated. A number of substances including those of endogenous nature discussed in Chapter **4** bind to $Na_V1.8$ channels by this mechanism. It can be assumed that direct interaction of an attacking molecule with the ion channel may not be perfectly specific: probably, the substance is capable of activating other channels which have homologous binding motifs in their structure. This may lead to adverse side effects at the organismal level. Nonspecific binding may even result in toxic effects upon application of the substance under investigation in the doses of the upper pharmacological range. We have chosen another approach. The strategy of development of novel analgesics should be based on the principle of physiological adequacy, which is illustrated by our studies on comenic acid effects. The agent specifically binds to the opioid-like receptor, the signal of which is strongly amplified by the transducer (Na^+,K^+-ATPase). As a result, the receptor-coupled modulation of $Na_V1.8$ channels leads to a decrease in impulse firing and, ultimately, to pain relief.

The authors believe that experimental verification of the assumptions discussed herein will not only deepen our knowledge about the mechanisms of nociception, but also will lead to the results of fundamental importance in clinical medicine.

CONFLICT OF INTEREST

The authors confirm that they have no conflict of interest to declare for this publication.

ACKNOWLEDGEMENTS

The authors are very thankful to Dr. Ivan Domnin for the synthesis and purification of comenic acid and other gamma-pyrone and gamma-pyridone derivatives. This book would not even exist without help of our late colleague Ms. Irina Katina. Her methodical approach and criticism ensured the reliability of our experimental data. We wish to express our gratitude to Prof., Dr. Andrey Derbenev for his outstanding findings at the starting part of our research and to Dr. Elena Karymova for her help with the experiments and mathematical modeling. The authors sincerely appreciate highly competent assistance of Dr. Svetlana Podzorova, her invaluable experience in the development and support of computer-aided systems formed the background of our experimental work. We are extremely grateful to Prof., Dr. Boris Zhorov for fruitful discussions. The authors would also like to thank Ms. Daria Semenova for graphic presentation of chemical structures and Dr. Valentina Penniyaynen for the cover picture of confocal image of neurite growth. Boris Krylov is cordially thankful to his dear teachers Prof. Dr. O.B. Ilyinsky and Prof. Dr. J.R. Schwarz. We are very much grateful to Michael W. Kenworthy, President of Technology Commercialization

Corporation, for his kind help and encouragement during research and development of novel non-opioid analgesic Anoceptin®. Investigations presented in this volume were funded by the Russian Science Foundation, research project No.14-15-00677.

GLOSSARY

Activation	A kinetic process of ion channel (voltage-dependent sodium channel herein) opening during membrane depolarization.
Action potential (nerve impulse, spike)	A transient voltage change recorded inside or close to a nerve, skeletal muscle, or other excitable cell, induced by a suprathreshold stimulating current. Impulse frequency resulting from generation of action potential trains is a manifestation of analog-impulse transformation produced by neural membrane. The rising (depolarizing) phase of action potential is due to functioning of voltage-dependent Na^+ and Ca^{2+} membrane ion channels allowing passive influx of Na^+ or Ca^{2+}. The falling phase is provided mainly by the passive K^+ efflux through voltage-dependent K^+ and leakage channels and due to inactivation of Na^+ or Ca^{2+} channels. Sodium spikes are brief (ca. 1-2 ms long); calcium spikes are more prolonged.
Active site	The part of a channel or any other cellular protein structure to which a substrate is bound.
Adaptation	A decline of impulse frequency response while a stimulus is maintained constant after onset. Spike frequency adaptation is manifested by an increase in interspike intervals while the stimulating current remains constant. Adaptation is an important characteristic of impulse coding.
Afferent nerve fiber	The nerve fiber (dendrite) of an afferent neuron (sensory neuron). It is a long process (projection) extending far from the nerve cell body that conducts nerve impulses from sensory receptors or sense organs toward the central nervous system. The opposite direction of neural activity is efferent conduction.
Affinity	The strength of binding of an investigated substance (ligand) to a membrane receptor, an ion channel, or another target. Affinity of ligand-receptor binding may be expressed by the ligand concentration at which half of the receptors are coupled with ligand molecules (the dissociation constant for the complex: K_d).
Anaesthesia	The absence of sensation. This may be general anaesthesia in which case the subject is unconscious, or local anaesthesia affecting the sensations from just a specific part of the body. Local anaesthesia may be due to influence of an anaesthetic drug, nerve injury, *etc.*
Analgesia	The absence of pain sensation.
Analgesic	A drug used to achieve analgesia relief from pain. Analgesic drugs act in various ways on the peripheral and central nervous systems. They are distinct from anesthetics, which temporarily affect, and in some instances completely eliminate, sensation. The most important manifestation of the action of an analgesic is a decrease of the frequency of impulse firing in afferent fibers.

Cardiotonic steroids	Cyclopentanophenanthrenes with a 5- or 6-membered lactone ring attached at the 17-position and sugars attached at the 3-position derived from plants which have long been used for treatment of congestive heart failure. They increase the force of cardiac contraction without significantly affecting other parameters, but are very toxic at larger doses. Their mechanism of action usually involves inhibition of Na^+,K^+-ATPase and they are often used in cell biological studies for that purpose. A fundamentally new role in nociception of two endogenous cardiotonic steroids, ouabain and marinobufagenin, is described in Chapter **4**.
Chelate complexes	Chelate complexes are formed by chelation. Chelation is a type of bonding of ions and molecules to metal ions. It involves the formation or presence of two or more separate coordination bonds between a polydentate (multiple bonded) ligand and a single central atom. Usually these ligands are organic compounds called chelants, chelators, chelating agents, or sequestering agents.
Chronic	Continuous or existing for a long time.
Control	Physiological systems are "control" systems if they are investigated at physiologically adequate conditions. They maintain homeostasis of some tissue and molecular parameters. A control experiment or "control" for short, means "check". Control experiments are designed to check that the result of an "effect" experiment is not due to an incidental aspect of one's protocol. For example, if one measures such a parameter as effective charge after drug administration, some prior "controls" should be performed to determine this parameter without the drug applied. After both "control" and "effect" experiments are carried out, the further analysis would elucidate the statistical significance of the difference between experimental and control data. This is a productive approach to verify the significance of the results obtained.
Drug	A chemical that affects biological tissues. The usage of this term is herein restricted to substances that are used clinically or for research purposes.
Electrical excitation	The phenomenon results from application of a suprathreshold electrical current which evokes a response in particular, an action potential generated by nerve or muscle cells (electrically excitable cells).
Frequency coding	A stronger intensity of the stimulus results in an increase of impulse frequency generated by excitable membrane. Spike frequency adaptation, numerical coding and frequency coding are under effective control of $Na_V1.8$ channels in nociceptive neuron. Novel analgesics should perform a physiologically adequate modulation of the gating device of $Na_V1.8$ channel responsible for coding of nociceptive signals, which is manifested in the decrease of impulse firing.
Inactivation	A kinetic process manifesting in the decrease of ion channel current switched on simultaneously with activation process by the stimulating current.
Ion channel	A protein that spans the cell membrane and allows passive movement of specific ions across the membrane. Many types of channels are gated: they can be opened or closed, *e.g.*, as a result of binding of various transmitters, hormones, or intracellular messengers, or as a result of changes of the membrane potential. Different channels also have different selectivities for the ions they let through. Single isolated ion channels can be studied by the patch-clamp method. Molecular mechanisms of modulation of ion channels form the basis for the functioning of the nociceptive system.

Membrane receptors	Proteins at the surface of a cell (built into its cell membrane) that trigger the cell signaling by receiving (interacting with) extracellular molecules. They are specialized integral membrane proteins that allow communication between the cell and surroundings. The extracellular molecules may be hormones, neurotransmitters, cytokines, growth factors, cell adhesion molecules, or nutrients; they interact with the receptor to induce changes in the metabolism and activity of a cell. In the process of signal transduction, ligand-receptor binding affects "descending" intracellular signaling processes resulting in gene expression.
Marinobufagenin	A cardiotonic bufadienolide steroid secreted by the toad *Bufo rubescens* and other related species such as *Bufo marinus*. It is a vasoconstrictor with effects similar to digitalis.
Na⁺,K⁺-ATPase	Sodium-potassium adenosine triphosphatase also known as the Na^+/K^+ pump or sodium-potassium pump, is an enzyme (an electrogenic transmembrane ATPase) found in the plasma membrane of all animal cells. The Na⁺,K⁺-ATPase enzyme is a molecular device that pumps sodium out of cells while pumping potassium into cells. This pumping is active (*i.e.*, it uses energy from ATP) and is of greatest importance for cell functioning. Additional signal-transducing (non-pumping) function of Na⁺,K⁺-ATPase is involved in nociception.
Naltrexone and naloxone	Opiate antagonists which block opioid and opioid-like receptors and are used to treat opioid addiction.
Na$_V$1.8 channel	Sodium ion channel encoded by the SCN10A gene in humans. It plays a key role in coding of nociceptive signals and it is the main therapeutic target for the development of novel analgesics.
Nociceptive system	This system involves nociceptors with specific molecular properties for coding of noxious signals, ascending and descending spinal and supraspinal pathways, and several distributed brain regions that regulate the processing and integration of nociceptive information with signals of other sensory modalities. Nociception (from Latin *nocere* "to harm or hurt") provides information about dangerous or harmful stimuli. Nociceptive signals trigger a variety of physiological and behavioral responses and usually result in a subjective experience of pain.
Numerical coding	The greater number of action potentials is evoked by a higher strength of the maintained stimulus (stimulating current). This phenomenon is an important manifestation of nociceptive information coding.
Opioid	An agent that binds to opioid receptors, located mainly in the central nervous system and gastrointestinal tract. There are four broad classes of opioids: endogenous opioid peptides, produced in the body; opium alkaloids, such as morphine (the prototypical opioid) and codeine; semi-synthetic opioids such as heroin and oxycodone; and fully synthetic opioids such as pethidine and methadone that have structures unrelated to the opium alkaloids. Although the term "opiate" is often used as a synonym for "opioid", it is more properly limited to the natural opium alkaloids and the semi-synthetics derived from them.

Opioid-like receptors	Membrane receptors which characteristics and physiological role is described in this volume. They are found in the peripheral nervous system and coupled to $Na_V 1.8$ channels through Na^+,K^+-ATPase which realizes its additional signal-transducing (non-pumping) function upon their activation. Opioid-like receptors are specifically activated by several gamma-pyrone and gamma-pyridone derivatives, as well as by morphine. Naltrexone and naloxone are the blockers of opioid-like receptors. Their molecular structure remains to be elucidated.
Ouabain	A plant-derived toxic substance that was traditionally used as an arrow poison in eastern Africa for both hunting and warfare. Ouabain is a cardiac glycoside and, in lower doses, it can be used medically to treat hypotension and some arrhythmias.
Pyridones	Isomeric derivatives of pyridine having a carbonyl group either ortho- or para- to the nitrogen atom. There are two isomers: alpha-isomer called also 2-pyridone, 2(1H)-pyridone and gamma-isomer called also 4-pyridone, 4(1H)-pyridone.
Pyrones	Pyrones or pyranones are a class of cyclic chemical compounds. They contain an unsaturated six-membered ring containing one oxygen atom and a ketone functional group. There are two isomers denoted as alpha-pyrone (2-pyrone) and gamma-pyrone (4-pyrone). The alpha-pyrone structure is found in nature as part of the coumarin ring system. Gamma-pyrone is found in some natural chemical compounds such as chromone, maltol, and kojic acid.
Signal transduction	Transmission of a molecular signal by triggering receptor structures coupled to intracellular protein complexes along a pathway terminating at the cell genome. Signal transduction occurs when a signaling molecule activates a specific receptor located on the cell surface or inside the cell. Depending on the cell, the response may alter the cell's metabolism, shape, gene expression, or ability to divide. The signal can be amplified at any step; thus, one signaling molecule can generate a response involving hundreds to millions of molecules.

REFERENCES

[1]   Pavlov IP. Conditioned reflexes: an investigation of the physiological activity of the cerebral cortex. London: Oxford University Press 1927.

[2]   Adrian ED, Zotterman Y. The impulses produced by sensory nerve endings: Part 3. Impulses set up by touch and pressure. J Physiol 1926; 61(4): 465-83.
[http://dx.doi.org/10.1113/jphysiol.1926.sp002308] [PMID: 16993807]

[3]   Mountcastle VB. The problem of sensing and the neural coding of sensory evens. In: Quarton GC, Melnechuk T, Schmitt FO, Eds. The Neurosciences. New York: Rockefeller University Press 1967; pp. 393-408.

[4]   Mountcastle VB. Central nervous mechanisms in mechanoreceptive sensibility. In: Darian-Smith I, Ed. Handbook of physiology, Set 1, The nervous system, Vol III, Sensory processes, Pt 2. Bethesda: MD: American Physiological Society 1984; pp. 789-878.

[5]   Krylov BV, Derbenev AV, Podzorova SA, Liudyno MI, Kuzmin AV, Izvarina NL. [Morphine decreases the voltage sensitivity of the slow sodium channels]. Ross Fiziol Zh Im I M Sechenova 1999; 85(2): 225-36.
[PMID: 10389179]

[6]   Katina IE, Shchegolev BF, Zadina JE, McKee ML, Krylov BV. Endomorphins inhibit currents through voltage-dependent sodium channels. Sensornye sistemi 2003; 17(1): 7-23.

[7]   Lopatina EV, Polyakov YuI. Synthetic analgesic Anoceptin: results of preclinical and clinical trials. Efferent Therapy 2011; 17(3): 79-81.

[8] Thirlwell MP, Sloan PA, Maroun JA, *et al.* Pharmacokinetics and clinical efficacy of oral morphine solution and controlled-release morphine tablets in cancer patients. Cancer 1989; 63(11) (Suppl.): 2275-83.
[http://dx.doi.org/10.1002/1097-0142(19890601)63:11<2275::AID-CNCR2820631136>3.0.CO;2-4]
[PMID: 2720576]

[9] Plakhova V, Rogachevsky I, Lopatina E, *et al.* A novel mechanism of modulation of slow sodium channels: from ligand-receptor interaction to design of an analgesic medicine. Act Nerv Super Rediviva 2014; 56(3-4): 55-64.

[10] O'Brien WJ, Lingrel JB, Wallick ET. Ouabain binding kinetics of the rat alpha two and alpha three isoforms of the sodium-potassium adenosine triphosphate. Arch Biochem Biophys 1994; 310(1): 32-9.
[http://dx.doi.org/10.1006/abbi.1994.1136] [PMID: 8161218]

[11] Lee CO. 200 years of digitalis: the emerging central role of the sodium ion in the control of cardiac force. Am J Physiol 1985; 249(5 Pt 1): C367-78.
[PMID: 2414999]

[12] Blaustein MP, Juhaszova M, Golovina VA. The cellular mechanism of action of cardiotonic steroids: a new hypothesis. Clin Exp Hypertens 1998; 20(5-6): 691-703.
[http://dx.doi.org/10.3109/10641969809053247] [PMID: 9682925]

[13] Wasserstrom JA, Aistrup GL. Digitalis: new actions for an old drug. Am J Physiol Heart Circ Physiol 2005; 289(5): H1781-93.
[http://dx.doi.org/10.1152/ajpheart.00707.2004] [PMID: 16219807]

[14] Lopatina EV, Yachnev IL, Penniyaynen VA, *et al.* Modulation of signal-transducing function of neuronal membrane Na^+,K^+-ATPase by endogenous ouabain and low-power infrared radiation leads to pain relief. Med Chem 2012; 8(1): 33-9.
[http://dx.doi.org/10.2174/157340612799278531] [PMID: 22420548]

[15] Yachnev IL, Plakhova VB, Podzorova SA, Shelykh TN, Rogachevsky IV, Krylov BV. Mechanism of pain relief by low-power infrared irradiation: ATP is an IR-target molecule in nociceptive neurons. Med Chem 2012; 8(1): 14-21.
[http://dx.doi.org/10.2174/157340612799278595] [PMID: 22420546]

[16] Krylov BV, Shchegolev BF. Substance with sedative effect. US Patent 7087640 B2, 2006.

[17] Krylov BV, Rogachevsky IV, Plakhova VB. Substance with sedative effect. US Patent 8476314 B2, 2013.

[18] Jarvis MF, Honore P, Shieh CC, *et al.* A-803467, a potent and selective $Na_v1.8$ sodium channel blocker, attenuates neuropathic and inflammatory pain in the rat. Proc Natl Acad Sci USA 2007; 104(20): 8520-5.
[http://dx.doi.org/10.1073/pnas.0611364104] [PMID: 17483457]

[19] Kort ME, Drizin I, Gregg RJ, *et al.* Discovery and biological evaluation of 5-aryl-2-furfuramides, potent and selective blockers of the $Na_v1.8$ sodium channel with efficacy in models of neuropathic and inflammatory pain. J Med Chem 2008; 51(3): 407-16.
[http://dx.doi.org/10.1021/jm070637u] [PMID: 18176998]

SUBJECT INDEX

www.ingramcontent.com/pod-product-compliance
Lightning Source LLC
Chambersburg PA
CBHW041727210326
41598CB00008B/803